CONFINED SPACE RESCUE

CONFINED SPACE RESCUE

Chase Sargent

Fire Engineering

PennWell

Published by Fire Engineering Books & Videos
A Division of PennWell Corporation
Park 80 West, Plaza 2
Saddle Brook, NJ 07663
United States of America

James J. Bacon, editor
Book and cover design by John Potter

Printed in the United States of America
13 14 15 16 17   6 5 4 3 2

Library of Congress Cataloging-in-Publication Data

Sargent, Chase, 1955-
    Confined-space rescue / Chase Sargent
        p. cm.
    ISBN 0-9212212-88-8 (hardcover)
    1. Industrial Safety. 2. Rescue work. 3. Industrial Accidents. I. Title.

T55 .S245 2000
628.9'2--dc21

00-026680

DEDICATION

For members of public services special operations teams, who, between responses, spend their entire careers training, studying, and preparing for the moment of truth.

ACKNOWLEDGMENTS

No part of this book would have been possible without the support of my wife, Kathy; my son, Brandon; and my parents, John and Gloria Sargent, all of whom have tolerated my erratic schedules, impromptu travel, and personal ravings for over seventeen years. My business partners, Mike Brown, Buddy Martinette, Jon Rigolo, Andy Spier, Dean Paderick, and Cindy Hendrick have endured my obstinate nature, and they have been persistently honest with me in return. Chief Harry Diezel (retired), Deputy Chief James Carter, and my bosses at the Virginia Beach Fire Department provided the training ground on which I could experiment, grow, and practice my craft. I must also acknowledge those members of the teams with whom I have worked who have demonstrated their merit and integrity even under the worst of circumstances. From among them, I would mention Battalion Chief Steve Cover, Firefighter Jon Rigolo, Firefighter Ray Haring, Captain Dennis Keane, Firefighter Wayne Black, Firefighter Keene Black, Dr. Dave Cash, and Captain Terry McAndrews. There are countless others whom I haven't the space to mention by name, including members of the Virginia Beach Fire Department's Technical Rescue Team and Hazardous Materials Team, as well as members of the FEMA Urban Search and Rescue Task Force, VATF-2. Finally, I would be remiss if I didn't mention my friends and mentors who have guided me along the way. Jim Gargan, Jim Hone, Tim Gallagher, Steve Storment, Mike McGroarty, Tommy Carr, Keno Devarney, Bob Samuelson, Ray Downey, Alim Shariff, Jim Segerstrom, John Taggart, John O'Connell, and hundreds of others have conferred on me their unwavering friendship and trust.

TABLE OF CONTENTS

FOREWORD

In 1986, Captain Robert Hoflin of the Little Creek Fire Department became one of the many statistics associated with confined-space entry and rescue operations. His death affected all of those connected with the team. Somehow we all felt partially responsible for his misfortune, for one can never know whether one more training exercise might have made the difference or not.

This book hasn't been born out of just that one experience, but also from more than two decades of both teaching and being a student of technical rescue. Although confined-space rescue operations represent only one of the many disciplines within the category of technical rescue, these operations also present the highest number of fatalities. Over the years, many outstanding courses and syllabi have been developed to teach the procedures involved. The instructors of these courses must be given credit for having saved an incalculable number of personnel from injury, incapacity, and death. Still, there has never been a single source of information that one could use to help train, equip, and operate a team in a confined-space environment. This book is the result of information compiled from a variety of sources, ranging from the efforts of some very wise instructors to my own experiences, and it aspires to be the first true text on these operations.

Part of the impetus for this book can also be ascribed to the changing environment of rescue services. The Occupational Safety and Health Administration (OSHA) wrestled with protective standards for years. Before standards were finally established, rescuers did the best they could. Sadly, often their efforts weren't good enough. With the development of OSHA 1910.146, *Permit Required Confined Spaces for General Industry,* the fire service, rescue teams, and safety officers received an eye-opening set of criteria for these operations. Although the standard is certainly neither definitive nor gospel, it is a step in the right direction when it comes to confined spaces and their associated problems. As OSHA faces reform, it may well be incumbent on rescue organizations to adopt its procedures voluntarily.

Although a text of this sort is intended to bridge the gap between theory and practical application, it is impossible for a single work to contain every last bit of information about a given subject. If you want to know absolutely everything about rigging and rope work, you'll really need to refer to a rope text. If you want to understand incident command down to its finest detail, you'll have to pick up the NFPA standard or perhaps one of a dozen books available out there in the marketplace. A full mastery of OSHA 1910.146 requires that one read the standard from cover to cover. The purpose of this book is not to delve into any of its subtopics to the deepest level, but rather, to lace them together at the top. Many of my colleagues and I were developing and performing technical rescue before it was in vogue, and that in itself connotes a responsibility to pass along what we have learned. I am happy to say that we have never had a team member killed or seriously injured during a confined-space response, or during any other technical rescue operation, for that matter. Hopefully, in the court of professional opinion, this will indicate that we are doing a few things right.

TOWARD FORMING A CONFINED-SPACE TEAM

*W*hile making the switch from ship power to shore power, as is the procedure for ships in port, a fire occurred in an electrical cable. The local fire department responded, extinguished the fire, and returned to quarters.

The ship involved was a military landing ship dock, whose internal well deck was used to house amphibious assault craft. Underlying the deck was a fuel cofferdam containing 82-octane mo-gas, used to power USMC vehicles. Due to a failure in the system, about two hundred gallons had leaked outside the tank. A civilian contractor had been called in to clean it up. The operation would require placing a worker inside the cofferdam by way of a sixteen-inch deck plate. Working in contaminated atmospheres in breathing apparatus and crawling through sixteen-inch baffles was just another day at the office for him.

As is typical in the cleaning industry, the electrical compressor supplying air to the worker provided no backup air system, and the worker carried no pony

1

bottle with him when he entered the hole. He penetrated over forty feet to the source of the spill and began vacuuming.

At about the same time, the ship's engineer decided that, with the electrical cable repaired, the ship should once again be placed on shore power. In making the changeover, the power surge blew out the fuses on the compressor, leaving the worker in the cofferdam without air. No one knows exactly what happened next, except to say that the worker struggled for a few terror-stricken moments and then quickly succumbed to the fumes.

The local fire department was notified, and they quickly responded. The members were already familiar with the ship, for it was the same company that had extinguished a cable fire on it earlier that day.

As the fire captain and his company approached the well deck, they were advised of the situation. A decision was made to attempt a rescue operation using standard fire service protective gear and standard SCBA. The captain and a firefighter removed their SCBA from their backs and, carrying this apparatus before them, entered through the same opening that the worker had used. The standard fire service gear that they were wearing proved to be too bulky, and the captain began to discard his boots and other personal protective equipment as he moved along, following the contractor's air line. At approximately the forty-foot mark, the trail led downward for another fifteen feet, passing through a floor baffle. Looking in, the captain saw nothing as his low-air alarm began to sound. He signaled the other firefighter to return to the opening for an additional bottle. As the firefighter left him, the captain passed his SCBA through the opening. He lost hold of his breathing apparatus, however, and it fell away, ripping the mask off his face and *causing him to bang his head against steel. Disoriented and trapped in a toxic environment, he lapsed into unconsciousness in a matter of seconds.*

Hearing the clatter of the falling SCBA and the sound of escaping air, the firefighter returned to the captain's position. He made several attempts to pull the captain out and even tried to give him air through his own mask, but the space among the baffles was too confined to accomplish either. Feeling the effects of noxious vapors himself, the firefighter made the difficult decision that saved his life. He retreated and collapsed at the entrance-way, and was pulled unconscious from the hole by fellow firefighters on deck.

A technical rescue team was summoned. In-line SABAs, still in the box, were brought in by helicopter, assembled at the site, and deployed with the rescue team. The delay for this alone amounted to thirty minutes. Six hours later, the body of the fire captain and the cleanup worker were removed.

Technical rescue has become almost a fad among fire, rescue, and industrial safety personnel over the past several years. I remember a time around 1984 when a group of us were practicing and refining technical rescue techniques, arguing over what the discipline should be called. Since then, the haz mat rage of the '80s has become just the day-to-day business of departments every-where, and the drift toward firefighter EMS has become an established fact. Despite these and other changes along the way, organizations continue to seek even more outlets for the services they provide, increasing their role in the community. Technical rescue is a most likely avenue for them to explore.

With this trend has come an explosion of private companies willing to teach technical rescue for a price. In many instances, the instructors have never even had any fire or rescue experience. Still,

Much of the technical rope work in use today was developed by mountaineers and wilderness search teams.

there's money to be made, and organizations are so hungry for information on technical rescue that they often jump at the first opportunity for training without even checking the credentials of those offering it. This is a lesson in itself. In the world of business (foreign territory to the average firefighter and other public servants), a low bid is a low bid, meaning buyer beware.

The art of technical rescue has a long but largely undocumented history. Well over a hundred years ago, the Fire Department of New York formed the Rescue Services Division. The teams within this division were intended to provide specialists for dangerous tasks, including the rescue of personnel who had become trapped or were overcome by smoke during fires. Much of the technical rope work in use today was actually developed by mountaineers and wilderness search teams. Within the rope community, there are still ongoing debates over various issues of techniques and equipment; whether a munter hitch is adequate; whether tandem prusiks are better than direct-loading cams; whether one anchoring system is superior to another. Typically, such debate with a given field spawns disparate cultures of experts squaring off against each others' techniques and philosophical approach. Out of such controversy comes the continued refinement of the craft, ultimately benefiting the means and methods best suited for any given set of circumstances.

Over the years, there has been a tremendous amount of fragmentation within the technical-rescue community itself, and there has been no clear voice to guide it toward uniformity. The very concept of what a technical rescue team is all about hasn't even reached maturity within most organizations. Still, the exotic, increasingly common scenarios that prompt the use of such teams have made many organizations realize that they are inadequately prepared.

Along with recognition of the technical category of rescue has come an avalanche of federal regulations, NFPA standards, OSHA requirements, and a plethora of other legislative, legal, and moral obligations. Listed below are just a few of the laws, standards, codes, and protocols with which an incident commander of a technical operation must be concerned today.

- OSHA 1910.146, *Permit Required Confined Spaces for General Industry.*
- OSHA 1926.650, *Trench and Excavations.*
- NFPA 1561, *Standard on Fire Department Incident Management Systems.*
- NFPA 1470, *Standard on Search and Rescue Training for Structural Collapse.*
- NFPA 1500, *Fire Department Occupational Safety and Health Program.*
- OSHA SARA Title III, *Superfund Amendments and Authorization Act.*
- OSHA 1910.134, *Respiratory Protection.*
- NFPA 1983, *Life Safety Rope, Harnesses, and Hardware.*

- NFPA 1670, *Standard on Technical Rescue.*
- NFPA 1006, *Professional Competencies for Technical Rescue.*

For the purposes of this text, technical rescue shall be defined as those areas of operation in which special equipment and techniques, as well as specialty personnel, are required to remove a victim from unusual circumstances or a unique environment. Some would argue that such a definition might encompass an ordinary vehicle extrication; however, vehicle extrications by hydraulic means have become commonplace enough so as to be considered the norm, almost routine. Rescues from trains, airliners, and ships, on the other hand, almost certainly require the intervention of teams with specialized technical skills.

It's true that confined-space rescue represents only one facet of a larger discipline, but many of its concepts and procedures are generic to technical operations as a whole. Any worthy discussion of one form will involve acknowledgement of the rest, since it is advantageous to understand some of the threads that tie these kinds of incidents together. Operational needs may vary from region to region, but it is enlightening to note how universal many of these technical skills are. With the possible exception of ice rescue, the following set of operational modes will find application virtually anywhere in the country, and the list is by no means exhaustive.

- Rope rescue operations, both high- and low-angle.
- Structural collapse operations.
- Confined-space rescue operations.
- Trench operations.
- Diving operations.
- Swift-water operations.
- Ice rescue operations.

- Tree rescue operations.
- Tactical helicopter operations.

All of these specialties require personnel, equipment, and training not traditionally mandated by conventional fire and rescue services.

Noting the commonalities of related disciplines introduces the dangers of trying to crossbreed them. Too often, especially where confined-space operations are concerned, an organization's official attitude is that "Special-equipping beyond a certain point becomes overkill," or "We use SCBA because we can't afford the supplied-air breathing apparatus systems." To the true professional, such an attitude is frustrating, almost laughable. In forming a dive team, no sane team leader would equip his members with SCBA in lieu of spending department money to provide them with scuba gear. If you wanted to form a haz mat team capable of Level A entries, you wouldn't hand them turnout gear just because the proper encapsulating suits would be too expensive. Such ideas sound patently ludicrous, yet the same sort of illogic is often applied to confined-space operations by many organizations, instructors, and teaching entities.

The demands of technical operations are such that would-be practitioners must pay full attention to the three key components of any such endeavor: personnel, equipment, and training. The personnel whom you choose for any special team must be special people—intelligent, fit individuals who can both think for themselves and operate as team players. The equipment that they receive must be designed and appropriate for the missions that they are to fulfill. Their training must be both foundational at the outset and comprehensive over the long term, inclusive of theory, knowledge, and practical application.

Even so, the actual process of building a team is a complex one, consisting of many aspects, some tangible, some intangible. Many times, we can identify key processes or behaviors that can indicate whether a team is meeting its goals. Moving along the continuum from "entry level" to "highly skilled" takes time, patience, and an understanding of the signs along the way. Establishing practical standards (i.e., measurable criteria) provides a good way of gauging whether individual members have achieved a projected level of competence or not. At the same time, certain abstract qualities, such as judgment under pressure and cooperativeness within the team, are harder to assess and defy direct measurement. The lessons of special operations teams within the military would indicate that such qualities are as vital to mission success as technical proficiency. Although the analogies drawn between war and civilian rescue operations are often overstated, it is true that a breach of team integrity or fault in judgment can have dire consequences in either venue.

Organizations expand into technical operations for a variety of reasons. In some instances, it is simply good politics to take a proactive approach to a potential or perceived problem within the community. Although decisions prompted by political expediency are usually suspect, this sort of department presents the easiest organizational climate in which to work. Don't become complacent if you are fortunate enough to find yourself in this sort of a department, for even the best of organizations can eventually be undermined by turf battles, cultural changes, and allocation issues driven by both internal and external forces.

Some organizations resort to technical operations in reaction to a mishap of some sort, typically when a lack of preparedness or other deficiency leads to a fatality. Although it may sound cold-hearted in the wake of a tragedy, this type of occurrence should be exploited to the net gain of the organization. Whatever the root cause of the original deficiency, the cure will almost certainly begin at the political and organizational levels. Change that comes about only after a calamity has occurred, of course, signifies a reactionary mentality on the part of management, and it is unfortunate that so many organizations are run this way.

The voice of labor often plays a dominant role in whether a department develops a technical capability or not, and that voice may be either pro or con, depending on local circumstances. Often labor will campaign for the development of a new service, and the reasons may range from preparedness concerns to job security to increased pay. At the other end of the spectrum, the union's position may be that its members shouldn't be required to master any new skill that falls outside of the traditional job description. The true motives of both labor and management in such situations often become obscured behind rhetoric and the ostensible reasons that they advance. Unfortunately, when a call comes in, notifying rescuers that someone is caught in a confined space, it is inconceivable that the fire department or industrial brigade won't respond. Invariably, personnel will deploy to the emergency environment, even if the members sent in are wholly untrained for that sort of incident. They will do something, anything, even if it's wrong. Sometimes they're lucky, and sometimes they're not.

Finally, there are those forward-thinking organizations that are already well ahead of the power curve. Driven by a specific geographical or industrial need, they prepare themselves at the outset for the call that will inevitably someday come in. These organizations don't act merely for political appeasement or in reaction to a tragedy; they don't need an outside agency such as OSHA to come in and tell them that they have a problem. They understand that, as public servants, their role is to meet the needs of the community, and they bother to take the time to assess what those needs might be. They have the willpower to work out any allocation issues along the way, and they seek the best resources for training and equipping their technical teams well in advance of any emergency.

In the aftermath of the incident described at the beginning of this chapter, several significant events took place. Within months after the accident, every military fire service was issued supplied-air breathing apparatus, something they had requested for years. Funds for confined-space training were forthcoming, and standards and operational guidelines for military installations were established. In the civilian quarter, supplied-air breathing apparatus suddenly became a priority for the technical rescue team, and it was rushed into service, along with the appropriate training.

Any number of factors contributed to that accident. From the outset, there was a failure to recognize in scale the "big three" requirements of the incident: training, personnel, and equipment. There was an obvious lack of training in hazard identification and confined spaces, coupled with a failure to select the appropriate personnel, followed by a failure to provide the proper equipment to accomplish the task. Relying on standard fire service SCBA was a horrifying error, and the removal of it from the users' backs so that they could pass through the tiny spaces only made the decision worse. Equally appalling was the lack of appropriate personal protective equipment. On a management level, the failure to implement the incident management system during the initial phases of the response, or a special-operations IMS later on, traces to a lack of standard operating procedures relevant to confined spaces. The department had also failed to provide Awareness-level training to all of its front-line personnel for these sorts of incidents.

Individuals often risk life and limb under dire circumstances to perform heroic deeds, but ultimately it must be realized that a confined-space operation is predicated more on team effort than individual bravery. The checks and balances of a competent, prepared team would almost certainly have prevented that fire captain from ever venturing below deck, at least in the manner he went. A tremendous amount of research has been done on teams, training, and organizations, much of it geared toward evaluating the effectiveness of military teams. Still, both military and civilian teams share some common denominators. Even when everything is being done correctly, rescuers who perform exceptionally complex and dangerous tasks require an extraordinary support network if they are to remain efficient. The premise of the team efficiency concept, or TEC, developed by Captain Mike Brown of the Virginia Beach Fire Department, is that safety, efficiency, and team integrity are synonymous. Training aimed specifically at fine-tuning team proficiency helps to uncover underlying deficiencies. By identifying each member's strengths and weaknesses, the team is able to compensate for and adapt to a range of changing circumstances

Training aimed specifically at fine-tuning team proficiency helps to uncover underlying deficiencies.

during an incident. At another level, however, TEC is concerned not so much with operational skills as it is with evaluating some of those more abstract qualities alluded to above: How does the team think? Do the members interact appropriately? Do they communicate effectively? Do they trust one another? Does one member recognize when another member is overwhelmed? Does he step in to help?

The issue of team efficiency is critical to an organization, yet it is often an overlooked facet of the entire program. Too often, an organization simply identifies a problem, then throws some money at it in the guise of training and equipment, hoping that that will solve the problem. What this sort of department winds up with is a short-term solution to a long-term problem. The effectiveness of an operation doesn't hinge on whether all of its members are skilled at tying knots. The crucial point is whether they are able to work in a concerted manner to carry out a difficult assignment. Inevitably, an organization that doesn't address this subtle but most vital aspect of operations will have trouble remaining fully functional, and it will be prone to injuries, flawed operations, and the premature burnout of personnel.

Unquestionably, technical rescue operations require personnel who are particularly well suited to working in stressful, alien environments. Although traditional emergencies such as fire and EMS operations are stressful by their very nature, the frequency with which they occur makes all but the worst of them fairly predictable and routine. It is axiomatic that organizations allocate funds based on the frequency of the various types of calls that they receive. Since incidents requiring technical rescue are relatively rare events, organizations tend to devote less time and money preparing for them than they do for more routine responses, with a consequent effect on a team's capability, performance, risk factors, and overall chances for success. At the same time, however, the members of technical teams are usually more highly motivated individuals, displaying a greater enthusiasm for their particular specialty and superior team loyalty. Of those technical personnel who do burn out, rust out, or otherwise become disaffected from a team, it can be said that a prime cause stems from having to perform double duty, responding to traditional calls on a daily basis while being expected to remain proficient in an arcane, undersupported, underrecognized specialty.

In lieu of high call volume, the training afforded a technical team should be aimed in large part at uncovering disfunctionalities across the spectrum of an operation. The ultimate goal in training for team efficiency is to create an awareness that no one individual can be highly proficient in every area of an operation. By identifying its inherent weaknesses, a team can compensate, adapt, and reinvent itself to best meet the requirements of the most unexpected call.

No matter how good your team and the individuals assigned to it, those members are still only human beings. Few personnel on a special team will be expected to perform only confined-space operations. Even if they are, the skills required to mount such operations are multidisciplinary in scope and can be applied to other areas. It is to an organization's benefit to expect, encourage, or require each team member to select both a major and a minor field of study. This will result in a well-rounded team, or teams, with knowledge pools distributed throughout. By following this practice, a team will better be able to fill the gaps during planning, training, or actual emergencies, and it will have a stronger hand in emergencies that involve related disciplines.

Finally, anyone who has spent any time in municipal or private rescue services is painfully aware of the political nature of the beast. The issues range from personal egos to turf to funding to competing programs to favoritism and favors, and back round to egos again. In paid departments, the cry is all too often "It's not my job." In volunteer departments, the common complaint is that "I'm just a volunteer—you can't expect me to take on all sorts of special responsibilities." No matter the root cause, politics will affect your ability to function. For a team to stick its head in the sand and ignore the political realities dooms it to a struggle from the very beginning.

Despite any perceptions to the contrary, the politics of technical rescue are an ever-present facet of team life within an organization. A successful team uses the resources at its disposal. It gains notoriety and ultimately attracts the most skilled, most motivated individuals. It has an easier time receiving allocations for its continued survival and health.

Keeping this in mind, every time you have an operation, whether for training or an actual response, be sure to get the media involved. Confined-space rescues are big news, regardless of how simple or complex a given incident may be from your perspective. Bring the media in and let them publicize the operation. Your supervisors, fire chiefs, plant managers, city administrators, and politicians read the same papers and watch the same news channels that you do. They will remember when it comes time to ask for a continued allocation of resources. Build a constituency that will support the team. Let the citizenry know that, although they may never have realized it, they just cannot live without the technical services that you provide.

Early on in the development of your team, and all throughout its operational years, adopt a sunshine policy. Let everyone see that the team is not some secret sect, but rather, that the input of the community is important, and that you expect their input and support. Always present both sides of the issue. Don't ever hold back the negatives associated with any program or team. Every program has options, not all of them good. You must be honest in your approach to dealing with management. Your job isn't to withhold data or information. Your job is to sell your side—out of your own experience and the knowledge you possess about a discipline that others may only partially comprehend. You have been to the recesses of a confined space. They haven't.

CHAPTER QUESTIONS

1. True or false: Much of the technical rope work in use today was developed by mountaineers and wilderness search teams.

2. What NFPA document is concerned with life-safety rope, harnesses, and hardware?

3. What NFPA document sets standards on technical rescue?

4. What characterizes a technical operation?

5. According to the text, what is a good way of gauging whether individual members have achieved a projected level of competence or not?

6. What is the mistake that an organization makes in avoiding the formation of a technical-rescue capability?

7. A confined-space operation is predicated more on _____ than individual bravery.

8. The team efficiency concept is concerned not so much with operational skills as it is with evaluating _____.

9. Since incidents requiring technical rescue are relatively rare events, organizations tend to devote less time and money in preparing for them, with a consequent effect on a team's _____, _____, _____, and _____.

10. Why should an organization actively bring in the media to let them publicize technical operations?

CHAPTER TWO

MANAGING THE RISKS

Whether your team is part of a municipal, industrial, or military organization, you will never operate under a more consistently dangerous set of circumstances than you will in a confined space.

Municipal governments and industry have entire staffs whose purpose is to manage risk. Most laws and consensus standards created for the fire and rescue community exist for the same purpose. Although risk management takes many forms, from keeping insurance costs under control to creating safe work environments, professional responders know that each day they go to work, they will take a risk. Fire, police, EMS, and industrial teams are paid to take a certain amount of risk that employees outside of our professions are neither required to take nor would ever imagine taking. For the responder, the question becomes how much risk is too much, and how can we intelligently manage the risks that we must take?

This is not a legal manual, nor can it replace good counsel from your city or corporate attorney. Still, as team members and leaders, we need to have a basic grasp of what risk management is all about. These are challenging times for those in the rescue community. Certainly everything that municipal responders do is under the microscope of public scrutiny. Many organizations face the loss of public funding, and in some cases a lack of confidence. Yet each time we respond, there is always a chance that something will go wrong. In reality, responders must practice risk management in everything that they do. As they say, if it's predictable, it's preventable. This text will attempt to describe the predictable aspects of a confined-space operation so that you can create the five pillars of success and risk reduction, namely people, policy, training, supervision, and discipline.

Why do things usually go right? In reality, operations usually go right, regardless of the complexity of the job, because personnel think quickly and draw on their experience and training. Of the two, experience is almost certainly the more vital factor. And why do things sometimes go wrong? Occasionally an operation will go wrong because of the personnel involved, but seldom does an organization or individual wind up in trouble out of malice. Actually, operations go wrong because individuals with worthy intentions find themselves in complex situations that evolve quickly, and in the span of one moment, someone makes a mistake.

Virtually all professional response personnel begin and end their careers with training. From a legal and risk-management standpoint, if personnel don't have much experience in handling confined-space incidents, due to low call volume and the relative rarity of these events, then they must rely solely on their training. If that training has been fairly limited, or if it hasn't been reinforced in some time, then the potential for mishandling the incident is far greater. Whenever a department neglects a given discipline because the call volume is low, it enhances the failure cycle.

By one definition, all responses can be grouped according to four basic categories: high risk/low frequency, high risk/high frequency, low risk/high frequency, and low risk/low frequency. Calls that present the worst potential are those in the first category, high risk/low frequency. Confined-space operations are unquestionably in this category. Personnel will simply not gain the experience or become as proficient at these operations as they will with structure fires, vehicle mishaps, food-on-the-stove calls, and brushfires. The only available method of managing the risks involved in a low-frequency event is to fund, train, and equip the team according to the hazards they might encounter, irrespective of any anticipated call volume.

Generically, there are many types of confined spaces, and an array of procedures that apply to them. The governing standard that this text will focus on is CFR 1910.146, *Permit Required Confined Spaces for General Industry*, since that is the one most applicable to rescue services.

Each year, according to OSHA, approximately fifty-four preventable deaths occur in confined spaces. The vast majority of these deaths, as well as injuries, are due to atmospheric problems within the space. According to the National Institute for Occupational Safety and Health (NIOSH), more than 35 percent of these deaths are among would-be rescuers, down from 60 percent before the standard. That puts emergency response providers from all sections of the community and industry in the group at greatest risk, especially first responders.

Every community has its share of confined spaces. Some common ones are tanks, silos, storage bins, storm culverts, hoppers, vaults, pits, trenches, railroad cars, and ships. Workers use manholes to enter sewers and utility vaults on a daily basis. OSHA believes that there are confined spaces in about 238,853 workplaces, and of the 12.2 million workers employed at these establishments, approximately 1.6 million make 4.8 million entries into these permit-required spaces each year.

Most confined spaces are entered infrequently for the inspection, cleaning, or repair of equipment. Some industrial areas, however, such as ships, aircraft, and storage tanks, are considered confined spaces during their construction, and worker entry during these periods may be frequent and routine. Many of the hazards

Every community has its share of confined spaces.

associated with the spaces described in this law may also be found in other technical rescue environments, such as collapsed structures, trench collapses, and caves, and although this standard wasn't written to address those areas specifically, the information and risk management plan outlined in 1910.146 offers valuable options when identifying and managing hazards in them.

The law was written for industry and then adopted by municipal, industrial, and military response teams as a template for risk management. It, along with others, was written for confined spaces such as those associated with agriculture, ship construction, and telecommunications, offering valuable information for rescue teams. Its most striking aspect is Section K, which outlines the responsibilities of rescue teams.

By definition, a confined space (1) is large enough and configured so that an employee can enter it, (2) has limited means of egress, and (3) is not designed for continuous occupancy. OSHA further distinguishes two specific types of confined spaces, non-permit and permit-required. Every confined space in an industrial setting is considered permit-required until an investigation by the employer reveals the nature and extent of its specific hazards. Nonpermit and permit-required spaces are thus distinguished by the hazards present and the ability of the employer to eliminate them. A space that does not require a permit is one that does not contain, or with respect to atmospheric hazards, has not the potential to contain, any hazard capable of causing death or serious physical harm. A permit-required confined space is one that displays one or more of several characteristics: (1) It contains or has a potential to contain a hazardous atmosphere, (2) it contains a material that could potentially engulf a worker, (3) it has an internal configuration that could trap or asphyxiate a worker due to its shape, or (4) it presents any other serious, recognized hazard.

Every confined space in an industrial setting is considered permit-required until an investigation by the employer proves it otherwise.

For the sake of argument, suppose that you are the risk manager of a major industrial producer of widgets. In the development of your confined-space program, as required by law, you begin to evaluate all of the spaces in your plant. You initially identify a given space as a permit-required space. Your original reason for classifying it as such is that it poses an atmospheric problem and has mechanical hazards such as mixing blades at the bottom. If you can successfully lock out the blades by cutting off their power supply, then you have successfully eliminated that hazard. The only remaining hazard is the atmospheric threat, caused by a residue that can be completely cleaned out, and the fumes from that residue can be removed by ventilation. Thus, you can document in writing that the space is hazard-free and will remain so as long as any worker is inside. If the cleaning procedures and other activities that occur outside of the space won't negatively affect the conditions within, the employer can classify the space as a nonpermit

If you can successfully lock out machinery by cutting off its power supply, then you can successfully eliminate that hazard.

confined space. There are a variety of other requirements involved in reclassifying a given space, of course, but this is the process in its simplest form.

The implication for rescue personnel is that any confined space to which we respond must be treated as a permit-required space, and we must consider the hazards identified above to be present until we can prove otherwise. Our immediate goal is to identify those hazards. In some instances, we may be able to eliminate all of them. In other instances, we won't be able to eliminate any. Our ultimate goal is to transform that space into a nonpermit space, making it as safe as possible for our team members. Unfortunately, being able to accomplish this is more the exception than the rule. Too often, we will have to operate under exactly the same conditions that the trapped, injured, or fallen worker encountered. From a risk management standpoint, you must base all of your tactical decisions on the assumption that the worst aspects of the environment exist until you can positively remove them.

In 1989, OSHA held public hearings on confined spaces and published four major findings.

1. There is a significant risk to workers who enter confined spaces.

2. Accident reports show a significant number of deaths and injuries resulting directly from entry into confined spaces. Because a sizable worker population enters these spaces, OSHA believes the risk to be excessive.

3. The risks associated with entering confined spaces include death, injury, permanent health impairment, and the loss of functional capacity. Most confined-space injuries and deaths result from asphyxiation due to hazardous atmospheres.

4. Many employers don't adequately protect their workers despite safety warnings.

The vast majority of mishaps occur because employees are poorly trained and equipped to handle the environment that they encounter. This also applies to would-be rescuers, who, according to accident reports, appear to be no better prepared or equipped than the original entrant.

Whether or not there is an OSHA standard covering your specific industry or occupation, workers are always protected by the general duty clause, {5(a)(1)} of the 1972 Occupational Safety and Health Act.

The OSHA standard for permit-required confined spaces is a performance-oriented standard. The agency does not intend to specify exact methods of compliance, provided that the employer's program results in worker performance that meets the requirements and prevents unsafe acts. Accordingly, there are fourteen areas of concern that must be addressed in every employer's permit-required program. Each employer must fulfill the following requirements.

1. Implement measures to prevent unauthorized entry.

2. Identify and evaluate permit-space hazards.

3. Develop procedures and practices for safe entry.

4. Provide, maintain, and ensure the proper use of equipment.

5. Test permit spaces for acceptable entry conditions.

6. Provide one or more attendants outside permit spaces.

7. Include procedures for emergency response by an attendant who monitors more than one space at a time.

8. Designate persons for entry operations, identify their duties, and provide them with training.

9. Develop procedures for providing and summoning authorized rescue and emergency services.

10. Implement a written system for preparing, issuing, using, and canceling entry permits.

11. Develop procedures to coordinate entry when the workers of more than one employer are simultaneously working in the same permit space.

12. Develop procedures concluding operations within the space.

13. Review entry operations and revise program deficiencies.

14. Review program and canceled permits annually for compliance.

As mentioned above, 1910.146 extends to the role of the emergency responder. No matter what sort of team you depend on or belong to, certain minimum requirements must be met to provide rescue in a confined space. OSHA allows the employer to choose the kind of rescue services that best fit a particular circumstance. The agency recognizes that some employers may not have the resources to maintain their own rescue and emergency services, but it also points out that prompt action by on-site responders may make the difference between life and death during a confined-space operation.

Four requirements in the OSHA standard apply to employers who have their own employees enter permit-required spaces to effect a rescue. Remember that these are minimums.

1. The employer must provide each member of the rescue service with the personal protective equipment and rescue equipment necessary for conducting operations within the space, and the employer must ensure that each member is properly trained to use that equipment.

2. Each member of the rescue service must be trained to perform all assigned rescue duties, and they must also successfully

complete the training required under paragraph (g) of the standard.

3. Each member of the rescue service must practice making permit-space rescues at least once every twelve months. Simulated rescue operations are to be conducted, during which rescuers must remove manikins or actual persons from the permit spaces or from a representative space. The representative space must have a configuration, a degree of accessibility, and openings that accurately resemble those of the actual space. If the rescuers perform an actual rescue in a satisfactory manner, this annual training is not required.

4. Each rescue service member must be trained in basic first aid and cardiopulmonary resuscitation. OSHA requires at least one member of the rescue service who has current certifications in both first aid and CPR to be available during entry operations.

It may be necessary for an employer to arrange for an outside contractor or municipal services to perform confined-space rescues. In choosing this option, OSHA requires the employer to inform the rescue service of the hazards that they may encounter when called on to perform a rescue at the facility. Additionally, the employer must provide the rescue service with prior access to all permit spaces from which a rescue might be accomplished so that the service can develop appropriate plans and training exercises.

In many instances, general industry simply assumes that the fire and rescue service in their area will respond and provide the necessary service when requested. Problems occur because, in many areas, fire and EMS teams simply aren't equipped, trained, or otherwise prepared to deal with the situations that they encounter. To address that issue, OSHA implemented a final rule in February 1999 updating 1910.146. The nonmandatory rule of Appendix F serves as a tool that can foster dialogue between employers and rescuers.

Beyond the requirements of OSHA, the fire and rescue services for years have lacked any national standard for technical rescue. The new NFPA 1670, *Standard on Operations and Training for Technical Rescue Incidents,* now provides an excellent platform from which organizations can mount these types of operations. It is a standard that codifies training requirements, organizational planning, definitions, and the caliber of operations. To complement this standard, another committee is working on NFPA 1006, *Professional Competencies for Responders to Technical Rescue Incidents.* That document will certainly mirror the requirements outlined in NFPA for the Technician level of responder.

Like most NFPA standards, 1670 provides the guidelines and general requirements that an organization needs to plan or provide technical rescue services. Operationally, it aligns itself with OSHA, ANSI (American National Standard Institute), and other national standards.

NFPA 1670 identifies seven specific technical-rescue disciplines that it addresses in detail, namely structural collapse; rope rescue; confined-space rescue; vehicle and machinery rescue; water rescue; wilderness search and rescue; and trench rescue. It also describes three levels of operational capability—Awareness, Operational, and Technician—and it outlines the training requirements for each. This allows an organization to pattern itself against a specific model of service delivery, and it allows individual members to alter their careers based on the level of capability that they achieve.

Like any NFPA standard, 1670 is a consensus standard. This means that there is no legal requirement for a department either to adhere to the criteria outlined by 1670 or to restructure programs to meet it. Almost every question about whether an organization should adhere to an NFPA standard is decided in court, after a mishap. The compliance of an organization hinges on whether it chooses to develop or reformat a curriculum that is specific to the standard, tailored to meet or exceed all of the criteria for that discipline at the appropriate level, whether Awareness, Operations, or Technician. Each organization must explore its options, examine its current programs, and develop strategies for becoming compliant; otherwise, the alternative is to decide that the organization will not provide standard-specific curriculums and will simply get on with business as usual. Many industrial brigades, for example, may have no need to maintain standard-specific training, and may instead only require training relevant to their particular environment.

As general requirements, the standard stipulates that each organization shall establish the level of capability that it intends to meet and that it develop standard operating procedures for that level of service. All members are to be trained at least to the Awareness level, and the organization shall provide training on a continuing basis to maintain the competence of the team. The authority having jurisdiction must document all training, and this documentation must be available for inspection by team members or their authorized representatives. Additionally, the organization must conduct a hazard and risk assessment of the response area to determine the feasibility of conducting technical rescue operations. Such an assessment includes the identification of potential hazards, plus the resources and procedures required to meet them. This assessment must be documented, reviewed, and updated on a continual basis. The jurisdiction is required to develop a process for planning incident response, meaning a formal, written special-operations plan. The plan must be distributed to all involved personnel and agencies. The team must be provided with the appropriate equipment, including personal protective equipment and breathing apparatus. As required by all operational NFPA documents, the organization shall ensure the safety of its technical rescue personnel through a variety of means, including but not limited to the use of safety officers and the incident management system.

These sections of the standard provide guidance for organizations in evaluating their current capability, and they provide a template for organizations just getting started in technical rescue. Following this process also enhances an organization's ability to make rational, fiscally sound decisions in evaluating what level of service to provide. For example, if a given response area is composed predominantly of ordinary, frame-and-masonry structures, the rescue team may only need to train to the Operations level for structural collapse to provide the necessary level of preparedness. Of course, the organization will need to train certain personnel to the Technician level in certain disciplines, but the decisions and resource allocations can be made by rational evaluation rather than just a gut feeling.

The three levels of capability are described as follows.

Awareness level: This represents the minimum level of capability necessary for a responder, in the course of his regular duties, to actively participate in a technical-rescue incident or take charge of the scene, as in the case of a first responder. Personnel at this level of training can perform search, rescue, and recovery operations. Members at this level are generally not considered rescuers, however. Awareness-level personnel may make assessments, make resource decisions, and perform some operations, but they are rarely involved in actual rescue operations involving entry into a technical rescue environment. In essence, this level of training offers some basic skills that a responder can use to evaluate and begin to control a technical rescue incident.

Operations level: This level of training focuses on the ability to recognize hazards, the use of equipment, and the knowledge necessary to support and participate in a technical-rescue operation. Search, rescue, and recovery may be involved, but these actions are usually carried out under the supervision of Technician-level personnel. Operations-level personnel can operate as team members, but they are limited with regard to the actual environment that they may enter, particularly with respect to confined-space rescue.

Technician level: This is the highest level of capability, focusing on the knowledge and skills required on the front line of an operation. Technician-level personnel are capable of performing, with few actual restrictions, in virtually any situation within their particular discipline. These personnel will make up the bulk of your technical rescue team.

An organization needs an effective blend of all three levels of personnel to deliver technical rescue services. As mentioned above, everyone in the organization is required to be trained to the Awareness level. A certain number of personnel also need to be trained to the Operations level so as to provide support and staffing at the incident scene. Specialty teams essentially need to be comprised of technicians. Naturally, how an organization achieves the right blend of personnel, training, and equipment is unique to each jurisdiction.

With the exception of rope rescue, all of the other disciplines defined by NFPA 1670 require composite knowledge and skills to pass from the Awareness level, through the Operations level, and on to Technician. Rope is actually the most fundamental of them,

Rope is the most fundamental skill required for technical rescue.

required for just about every other discipline as an entry-level or baseline skill. As an example of a career path, the prerequisites for obtaining a confined-space rescue technician rating include a confined-space and a rope-rescue course, both at the Awareness level. Additionally, the candidate must meet the requirements of NFPA 472, *Professional Competence for Responders to Hazardous Materials Incidents*. The completion of these courses will qualify him for certification at the Awareness level. For an Operations rating, the applicant must complete a rope-rescue and a trench-rescue course at that level. The final advancement to Technician is predicated on completion of a confined-space operations course, as well as a confined-space Technician course.

At the Awareness level, confined-space personnel are expected to perform an assessment of the site, recognize its hazards, determine the resources required, and keep themselves and others safe. Nonentry rescues are permitted. For example, an Awareness-level responder may extricate the victim using a long-handled tool, or perhaps haul him to safety using a winch or pulley system to which the victim is already attached. The Awareness-level responder is not allowed to enter the space, nor can he perform atmospheric monitoring or certain other functions associated with entry.

Operations personnel are expected to perform at a much higher level of capability. These personnel are expected to perform atmospheric detection and monitoring, as well as to interpret the results. Operations personnel may also perform entry operations; however, the parameters are quite clear, and certain restrictions are placed on team members rated at this level. Entry for these members is permitted if the circumstances meet these and other criteria.

1. The internal configuration of the space is clear and unobstructed, so that retrieval systems can be used for rescuers without the chance of entanglement.

2. The victim can easily be seen through the primary access opening to the space.

3. The rescuer can easily pass through the access opening with room to spare when wearing personal protective equipment in the manner recommended by the manufacturer.

4. The space can accommodate two or more rescuers, in addition to the victim.

5. All hazards in and around the space have been identified, isolated, and controlled.

Chapter 3 of NFPA 472 outlines the parameters for the Technician level. Organizations need to review this section carefully, since its requirements are quite substantial for responders.

NFPA 1670 offers unique challenges and opportunities for the fire service and other rescue organizations. Finally a document exists that promotes common approaches to technical rescue incidents. Like anything new, the standard creates some pitfalls for organizations as well, since it reintroduces such issues as training, funding, organizational commitment, and service delivery from a whole new perspective. Considerations related to the underlying values of an organization are often the most difficult barriers to overcome.

CHAPTER QUESTIONS

1. As described in the text, what are the five pillars of success and risk reduction?

2. According to the National Institute for Occupational Safety and Health, would-be rescuers currently account for approximately what percentage of fatalities in confined spaces?

3. By OSHA estimates, how many permit-required entries are made into confined spaces each year?

4. By definition, what is a confined space?

5. What is a nonpermit confined space?

6. Most confined-space injuries and deaths are the result of _____.

7. Why do the vast majority of confined-space mishaps occur?

8. Why is CFR 1910.146 considered a performance-oriented standard?

9. What is the importance of NFPA 1670?

10. What are the seven technical-rescue disciplines, as identified in NFPA 1670?

11. May confined-space personnel trained to the Awareness level perform nonentry rescues?

12. May confined-space personnel trained to the Awareness level monitor a space for atmospheric hazards?

13. Under what circumstances may Operations-level personnel enter a confined space to perform a rescue?

PLANNING

Every time we respond, we take a risk, a calculated one. And each time we successfully mitigate some form of emergency, we learn from it and incorporate those lessons into reducing our risk the next time out. Each time we respond to a certain type of emergency, the better and better we get at what we do.

Still, any response mode naturally brings along increased risk for an organization. There are three basic options from which a department can decide its ultimate future: It can refuse to respond to confined-space emergencies, it can function as a backup response team, or it can accept its responsibility and prepare. In following the first option, a department immediately makes itself vulnerable, for when the 911 line rings, it is inevitable that someone will be dispatched to the site. No group of firefighters, police, EMS responders, or personnel from other agencies are just going to stand idly by at the rim of a confined space and do nothing. How can a department possibly expect to screen a 911 call, regulate the response, and stop action from being taken? Your department may say that it intends to pursue a hands-off course of action—just explain how you'd intend to make it work!

By the same token, performing exclusively as a backup team may or may not be possible. If you respond to an industrial area that already has a fire brigade or rescue team, perhaps that other entity will act as the primary response agency and yours will only have to function in a support capacity. It is unlikely, however, that the only confined spaces in your jurisdiction are connected with industries that maintain prepared brigades. A municipal department will need to address all of the others, and with no one to play support to in those instances, its officers and personnel won't be left with much of a choice.

The only course that makes sense in the real world is to accept the responsibility and prepare. Once you have identified the confined spaces that pose a threat in your community, you will be compelled to prepare for them. That means determining what level of service your organization should provide, whether Awareness, Operations, or Technician. At the heart of realizing any practical capability is planning.

Planning can save time, money, and lives if it is done correctly. It may be impossible for a municipal agency such as a fire or EMS team to identify and plan for every confined space in the response area. Industrial teams can have a much easier time of this, since their response areas are limited to their worksites and, in many instances, they work in and around those confined spaces every day. Regardless of the environment in which your team operates, there are some simple rules to follow that will help you plan for confined spaces.

Preincident planning can be broken down into several phases.

1. Assessment of the hazards and risks.

2. Development of internal and external resources.

3. Assessment of team readiness.

4. Training of the team.

5. Recurrent training and periodic reassessment of the program.

In assessing the hazards and risks, it's necessary for the authority having jurisdiction to determine the following.

1. What confined spaces exist in the response area?

2. What are the hazards associated with each space?

3. Do any hazard spaces present a unique or complicated challenge?

4. What risk will the organization incur, and what level of risk is acceptable?

5. What can the organization do to reduce the risk?

6. At what level of capability does the organization need to function to respond to the identified sites?

7. Do any of the confined spaces present a hazardous materials problem?

Confined spaces exist all around us, but planning for them can be deceptive. Many municipal organizations won't fully understand the unique situations that they may encounter in an industrial setting. It isn't uncommon for a department to assess a bulk storage facility for the purposes of firefighting and completely ignore the area's potential for being the scene of a confined-space operation. Remember OSHA's mandate that a rescue team must be given prior access to permit-required spaces for the purposes of inspection and training. The term "industrial area" can mean anything from manufacturing plants and assembly lines to storage and shipping facilities to bank vaults and prisons to quarries and mines. When you go out to familiarize yourself with these areas, you will discover that many of them also present a unique hazard. An organization shouldn't be deceived just because the call volume to such locales may be low, for the accompanying operational risks remain astronomically high. The challenges may be a matter of access routes, water, security clearances, hazardous materials, spatial configuration, or all else; still, it is almost a certainty that a department that has identified these hazards and planned for them beforehand stands a better chance of surmounting the risks than one that hasn't.

It isn't uncommon for a department to assess a bulk storage facility for the purposes of fire-fighting and completely ignore its potential for being the scene of a confined-space operation.

Another task that a department faces is to assess its resources at hand, both internal and external. Falling under the category of internal resources, of course, are those that an organization holds in its own inventory, such as apparatus, tools, team-related equipment, breathing apparatus, communications equipment, special teams, and general staffing. External resources are the essential services or hardware that a department may need to acquire from an outside agency or vendor. Some examples of external resources include specialized equipment, industrial cleanup companies, marine chemists, laboratories, public utilities, the department of public works, private contractors, and heavy equipment.

One certain way to obtain many of the necessary resources is simply to budget for them. Obviously, if you have identified a problem and are expected to respond to it, there is always a cost attached. If you intend to make special acquisitions within the budget, remember that your target is to be in compliance with federal, state, and local mandates. If you're going to respond to confined spaces, you are required by law and consensus standards to meet certain minimum equipment and training standards. Make sure that the administrators and local politicians understand this up front. Document all of your requests, and be sure to keep a verifiable record of those that are rejected. If an incident occurs and your team doesn't have the right equipment, then the media, the citizens, the politicians, or the lawyers are going to eat you up. If you can show that you attempted to acquire those resources and were denied for budgetary reasons, then the burden of explanation or liability may well fall elsewhere.

Although industries are required to provide some form of confined-space rescue capability, whether internal or external, many simply won't address the issue until an incident occurs. Their reaction at that time may simply be to call the fire department. If yours is an industrial team, you may want to consider contracting your services out to other industries in your area.

If yours is a municipal team, you may be able to acquire resources through a confederation of industries within your jurisdiction. If a local industry intends to rely on the public emergency services, it is appropriate that they bear some of the costs. Individually or as a group, industries can purchase equipment that your team specifies; otherwise, they can provide the funds for it. Not only will this contribute to the team's capability, it will also immediately expand the arsenal of equipment that can be considered internal resources. We in municipal departments are all aware that it is easier to get an item that is already in inventory replaced than it is to requisition an item that the department has never owned before. Thus, once you put into service a piece of equipment that a local business has purchased for you, and once you put a city or plant ID number on that item, then the process of having to repair or replace it, using budgetary funds, should prove simpler than the original acquisition.

Another way of acquiring the resources you need is through local organizations. Consider the following.

1. National Safety Council.

2. Public utilities and the department of public works.

3. Utility companies.

4. Utility contractors associations.

5. Building contractors associations.

6. Nonprofit organizations dedicated to community services, such as the JCs and Kiwanis Club.

7. Other nonprofit organizations.

8. Telephone companies.

After you've identified these potential patrons, get to know them. In some instances, they may have useful equipment simply lying around in a back storage area somewhere. Our team has found supplied-air breathing apparatus, retrieval devices, and atmospheric monitors that certain organizations had purchased at one time and never learned to use. These items were donated to us for use in confined-space operations.

It may also be of benefit to a department to either form or attach itself to a nonprofit organization. My department works with the

Tidewater Emergency Medical Services Council, a nonprofit corporation. Our unique relationship serves both parties. The council is assured that it remains on the cutting edge of providing technical rescue services throughout the region. The benefit to the team is that it has an independent body keeping the books on any donations or income gained through instructional seminars and the like. When it comes to finances, it's best to have an independent agency facing audit rather than an individual or the team itself. This arrangement also provides a tax write-off opportunity for companies and individuals who donate to the team. Since TEMS is a nonprofit entity, we can give the donor a letter stating that their contribution was made to a charitable cause.

Despite earlier comments, make sure that you familiarize yourself with your municipality's policy regarding the acceptance of donated equipment and the process of placing it in inventory. Some municipal governments consider this to be a way of stacking the budget and therefore shun the practice. They know that, in the long run, they will end up replacing it. A team would still have the option of purchasing whatever it wants and using it for response; however, it would have to pay repair and replacement costs out of its nonprofit accounts.

As mentioned earlier, a serious organization will early on decide the level of service that it intends to provide. Naturally, citizens expect the highest level of service possible in the event of an emergency, and there are both legal and moral obligations regarding a department's concern for the health and safety of its employees. When it comes to confined-space operations, I would suggest that all jurisdictions have a need for an advanced capability, and if you were to review the restrictions placed on Operations-level personnel, you will likely conclude that Technician-level service is a must.

It is a given that all personnel within the organization must be capable of operating at the Awareness level of service. You should realize, too, that you will need to provide Technician-level responses and service at some time during your career, and that your ability to deliver service in the realm of confined-space rescue

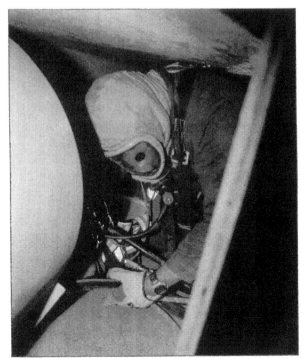

Given the restrictions placed on Operations-level personnel, a Technician-level capability is a must.

depends on that all-important mixture of personnel trained to the appropriate levels. Setting the personal requirements for membership on the team may sound like an easy question, but it is a matter that can become very political from time to time. Many team leaders, for example, meet with resistance from department brass when they try to establish strict qualifications for membership. Still, technical rescue teams should consist of top-gun personnel. These individuals should be highly motivated, highly trained, and highly capable. The surest way to destroy morale, eliminate team efficiency, and compromise safety is to let just about anyone join up, irrespective of their qualifications.

Consider the following criteria for inclusion on a team. None of them are carved in stone, so a given organization can adapt them as it sees fit.

1. Personnel should have amassed a minimum time in service, perhaps up to five years, before being allowed to join the team.

2. The selection of candidates should be made by a board. This would entail an interview process by officers and at least one senior member of the team.

3. Candidates should be evaluated based on their ability to function as a team member, as well as their current level of training. Preference should be given to those who've been motivated enough to seek out training on their own.

4. Selected members should be evaluated six months after joining the team and again at twelve months. Those who fail to make the grade should be removed by the end of the first year, regardless of their overall time in service or career experience.

5. Team members should be required to meet certain physical fitness standards, above and beyond those required for conventional fire service operations.

The qualifications of the team's leaders deserve equal scrutiny. Apparently, many organizations believe that the leadership of a team doesn't have to meet the same standards as its members. This is a mistake. In municipal organizations particularly, senior staff chiefs have been known to take command of special programs, even without displaying a hint of technical competence. Let it be said that organizations or individuals who believe that they can lead a special team based on their rank and not their technical expertise will fail. The members will soon lose all faith in them, and their commitment to the team will soon erode.

Whoever you choose to lead your team must have the same level of competence as the team's most effective member. Leaders must train, maintain all certifications, and be prepared to function as the operations officer at the vast majority of incidents. The distinction between functioning as an incident commander and an operations officer is an important one. At the scene of a confined-space emergency, the incident commander is usually a resource manager and planner, having limited training in confined-space operations. From the standpoint of the team, the operations officer is the real organizational hub, evaluating and implementing strategy and tactics at the rescue site.

A team at the Technician level should be stationed in one location, and its members should live, train, eat, sleep, and respond as a team. Departments with more than one team may opt to station them in different locations, but each one is a unit that should not be broken up. The squad concept works exceptionally well for municipal services. For industrial teams, squads can be composed of personnel assigned to the same work shift.

Similarly, an organization must decide how it will station its Operations-level personnel. My department currently uses ladder companies to support its special teams. In the industrial setting, it may be necessary to train all team members to the Technician level, since staffing may be more of an issue than it is with a municipal fire department. A large industrial complex may be able to develop an Operations-level team within the work groups, however.

Whether the team is municipal or industrial, response guidelines are mandatory. These need to be published, and all of the people within your organization must be made aware of them, including dispatchers and other communications-oriented personnel. The protocols that you set should be entered into your computer-aided dispatch (CAD) system to ensure that you'll get the resources you need, when and where you need them.

CHAPTER QUESTIONS

1. Why may it be infeasible for a municipal department to act only as a backup team to the industrial brigades in its jurisdiction?

2. Preincident planning can be broken down in what five phases?

3. True or false: It is an OSHA mandate that a rescue team be given prior access to permit-required spaces for the purposes of inspection and training.

4. In planning for a response capability, industrial cleanup companies, marine chemists, laboratories, public utilities, and private contractors may all be considered examples of _____.

5. When requesting special acquisitions within the budget, why should you document all of the requests and keep a verifiable record of the rejections?

6. Why may it be of benefit to a department either to form or attach itself to a nonprofit organization?

7. The team leader; i.e., the operations officer, should have the same level of competence as _____.

C H A P T E R   F O U R

TRAINING THE TEAM

A special-operations response team will spend far less time on actual calls than it will in training for them, but that time spent in preparation will make all the difference when the alarm does sound. Put simply, training is the essence of success. Any failure to train your personnel adequately will serve nothing but a false sense of security, and no one should venture thirty feet under a rubble pile or deep into a fuel tank without an appreciable understanding of the issues involved.

Naturally, the training within any organization should be formatted to meet the potential response needs. It should also be realistic and scenario-driven, conducted within real time frames. Training that is realistic will present a certain amount of risk, and since the experience gained may save someone's life in a real situation, the challenges should be met rather than watered down and sanitized.

Training can be conducted by in-house personnel; outside contractors; through state or local instructional institutions, such as community or technical colleges; or by implementing a train-the-trainer program at different levels of capability. Whatever path you choose, your instructors must be high-caliber personnel with adult-education skills, plus a significant background in technical-rescue operations, not just confined space. Someone who goes to a course or two and comes back an expert should be considered more of a liability than an asset.

If you hire a private company, make sure that you do your homework ahead of time and consider five key questions when choosing the company and its instructors.

1. Are the instructors and owners active in technical rescue or emergency services, or are they just out to make money?

2. Are the people you're hiring adult educators or merely instructors?

3. Will a company's low bid mean substandard quality?

4. Have the instructors ever responded in the real world, or has all their experience been gained in the classroom and simulated exercises?

5. Is the company insured?

Organizations can do much to develop evolutions and props at relatively low cost. Take a look at some of the modern playground equipment, including those colorful tunnel mazes found in certain fast-food restaurants. A similar assemblage of pipes, slides, and pits, rendered in PVC, wood, or steel, can easily serve as a confined-space mockup at any fire, industrial, or military training site. Aside from budget, the props and sets that you use are limited only by your imagination, ingenuity, and construction capabilities.

Whether a given training session is conducted in the classroom or in a practical environment, it should always have a secondary purpose. The objective above and beyond the ostensible purpose of

A portable training simulator made from steel pipe.

any exercise should be the development of the team. At every opportunity, we should be urging and expecting the members of a team to work more and more cooperatively. In the classroom, this may entail running a game exercise in which personnel work together to build a team as a functional unit, aimed at carrying out a task. The members should elect their own leadership and demand accountability from all involved, meaning each other. Simple exercises, such as having to remember everyone's name or lining up the members by height, can be used to help initiate the process. An instructor should always promote a healthy spirit of competition between the squads. One classic way to do this, of course, is to time a team as it performs an evolution, then challenge the next team to better it. Teams and their members should be encouraged to think creatively, "outside the box," so as to find better ways of accomplishing a given task. Even if they fail, the experiment will prove to be a learning experience, for it is better that they improvise and fail now than during a real operation. Hide-and-seek games involving communication, the effective transmittal of information, can also be valuable teamwork exercises. During class assembly, squads and teams should be encouraged to challenge each other. An instructor can even favor one team over another, subtly allowing them to gain a margin. This can drive the other team to do even better, thus stimulating competitive, cooperative behavior among the members.

Any basic course needs to begin in the classroom. All Awareness- and Operations-level classroom instruction must address the following areas, at minimum.

1. Why confined-space rescue must be considered a specialty operation.

2. A review of local, state, and federal laws governing confined-space operations.

3. Recognizing confined spaces.

4. The levels of training and response capabilities.

5. The phases of a confined-space operation.

6. Making fundamental determinations, such as those related to size-up, the risk-to-benefit analysis, operational mode (rescue versus recovery), and the like.

7. Recognizing and controlling hazards.

8. Understanding the need for, use, and limitations of equipment.

9. Selecting proper breathing apparatus.

10. Atmospheric monitoring.

11. Basic ventilation concepts.

12. Emergency escape procedures.

13. Personal protective equipment.

14. Safety.

15. The psychological component of confined-space operations.

16. Team concepts.

Technician-level instruction can include additional, more comprehensive classroom sessions on such topics as atmospheric monitoring, hazardous materials, planning, incident management, legal updates, team decision-making, advanced rope skills, marine operations, and other subjects as appropriate.

It is important that personnel become intimately familiar with any equipment that they may use. A workshop format is excellent for developing a baseline understanding of how a given tool works, how it's put together, what maintenance it requires. Workshops usually involve breaking up the class into squads and having them rotate through four or five stations, each with a different theme, but in reality all part of the same picture. An example of this sort of integrated workshop might include stations concerned with personal protective equipment; patient packaging and removal equipment; atmospheric monitoring and ventilation; supplied-air breathing apparatus; and support equipment. Workshops allow you to use technicians for instructional purposes, and they're great places to groom future instructors.

As mentioned above, practical training can be conducted in tunnels and mazes constructed specifically for functional exercises. I refer to training in these sorts of environments as having a nonmission profile; that is, the students practice in an environment that doesn't require them to use all of the equipment that they'd need in the real world. Training conducted with a nonmission profile is geared toward developing skills relevant to certain equipment only, and perhaps to introduce the student to a specific spatial configuration. These types of practical sessions really have four key purposes: (1) to foster teamwork, (2) to have a student work with a particular piece of equipment in a confined space, (3) to prepare the student psychologically for real-world confined spaces, and (4) to build on

An old airplane fuselage reconfigured inside to create a confined-space training simulator.

basic skills. Entry and tight-space scenarios can be created by using large sections of metal pipe configured in a variety of angles. Concrete and PVC pipe can also be used, as can plywood or a variety of other materials. Personnel must pass through spaces of different sizes and length, with or without personal protective equipment. The exercises may include long traverses in teams of two or more, or they may involve individuals moving through sections of pipe as narrow as fourteen inches or so.

Maze training is usually done in teams of two, and it is timed. Besides the lessons of teamwork, such an exercise can help a student learn to slow down in order to speed up. By pushing the students to a point where they are working so quickly that they become inefficient, the instructor can then slow them down again, demonstrating how they can actually increase their proficiency and cooperation. It's useful to burden the students with other tasks, such as having to manage SABA lines, as they deal with the exigencies of the maze itself.

One exercise directly involving SABA requires students, paired off in teams of two, to don the apparatus and maneuver as a team

The FDNY house maze provides a variety of internal configurations guaranteed to challenge a student.

past an array of obstacles. The obstacles could be as simple as tables and chairs placed on their sides and stacked in a classroom, or they could be something infinitely more complex. The students should be asked to make their way through the course and then back out again. At the turnaround point, the team is usually required to package a patient in a transportation device and then drag him out. Such an exercise involves a wide range of skills and cooperative effort, and the students can become acclimated to SABA in a controlled environment. Other team members outside the maze must monitor the air-supply system, tend the lines, or handle communications. All of these are skills that each team member will have to master.

Another exercise requires the student to use SCBA in a variety of timed evolutions, such as assembling, donning, and doffing the apparatus. This instruction should also include the firefighter-down position, which is the position that a rescuer takes when he is trapped and awaiting assistance. Lying on your side with your legs extended and your head on your outstretched arm allows you to conserve air and protect yourself. Other skills, such as communications and line tending, can be incorporated into this exercise. Many students take SCBA almost for granted, but this sort of instruction can enhance the skills that they already have.

In an in-and-out drill, team members are required to place teams of two inside a vertical space, then remove them using a variety of methods. This exercise entails using rope skills, including mechanical advantage systems and lowering systems. It familiarizes them with tripods, ladder derricks, hoists, and a variety of other tools, and it also teaches vertical line tending skills, which are critical.

All of the training done in classrooms, workshops, and practical scenarios allows a team member to build on the skills he has previously learned. Each member must work in a controlled environment,

In-and-out drills increase the proficiency of students in a number of hoisting techniques.

improving his techniques and ability with equipment. These are building-block exercises, getting personnel ready to put the pieces together in an actual scenario. In an exercise with a full-mission profile, teams must operate exactly as they would in the field, using all of the equipment and other resources at their disposal. There are two types of full-mission-profile exercises, managed and facilitated.

A managed full-mission-profile exercise is an excellent starting point for consolidating all of the skills that the team has been practicing. It is managed in that an instructor or senior team member actively coaches the operation, making suggestions and providing constant feedback. In effect, the instructor plays an active role in the outcome, for his comments help shape the operation as he guides the students toward proper techniques and appropriate decisions.

Managed profiles offer the most effective way of inducing and evaluating the processes inherent to decision making within the team. No team should be allowed to advance into facilitated scenarios until its members have run numerous managed scenarios. Personnel who engage in high-risk specialties must be able to identify and make key decisions about operational issues, adjusting their tactics as necessary. The only way for them to become proficient in these skills is to have someone manage the process and provide feedback.

Once a team has shown that it can function at the managed level, its members are ready to advance to a facilitated full-mission-profile exercise, meaning that the entire mission is run without interference from the facilitator. No feedback is provided until the end, and then only in the form of a critique. The facilitator sets up the scenario, lays down whatever rules may apply, and in some instances may trigger certain complications during the evolution itself, such as a partial collapse or a simulated loss of electricity. Unless there is a critical safety issue that requires that the operation be stopped, reevaluated, or redirected, the facilitator does not interfere. A facilitated profile can be an excellent training tool, although an instructor must understand exactly what is gained by this approach. It is best suited to advanced teams that can recognize and modify their methods during the course of a mission. Since instant feedback is not provided, its value as a tool for team decision training is limited. Its true worth lies in approximating the context of real-world operations and their associated problems.

In this scenario, students using SCBA and hardwired communications systems negotiate obstacles to reach and package a patient.

Even though training may be conducted under the auspices of a full-mission-profile exercise, it is important to control the environment in terms of atmospheres, chemicals, machinery, and utilities. Every space should be devoid of imminently dangerous atmospheres prior to entry, no residual chemicals should exist, and in no instance should you ever place trainees in areas containing live hazards, such as steam, electricity, and the like. The presence of haz mats can always be simulated using nontoxic smoke.

A good example of an exercise with a full-mission profile is one involving a tank truck. In this exercise, a victim is placed inside a 406 tanker, similar to a gasoline tanker, and personnel are required to perform a full mission to rescue him. The tanker can be dry or wet, about half-filled with water. Running the exercise wet makes movement between the baffles more difficult. It heightens phobic reactions and generally makes life for the rescue teams miserable. All of this is for the best, since team members are going to be miserable in the field. Cold and wet is a way of life in technical rescue.

An exercise with a full-mission profile conducted in a vault or sewer involves deploying a team through a manhole and inducing the members to move through the interconnecting pipe system to reach a victim in another vault or pipe. This exercise can also be run wet or dry. This is a great evolution, since utility vaults and sewers are common to all communities.

Exercises geared for industrial settings can be conducted in towers, reaction vessels, grain silos, railroad hopper cars, ovens, storage facilities, and other likely locations. If there is a maritime presence in your jurisdiction, you must train in these locations as well. Ships, shipyards, and dry docks offer many possibilities for full-mission-profile scenarios. Bulk-storage facilities, such as floating-roof oil tanks, also offer unique circumstances. Virtually any structure with height and narrow routes of access will warrant inclusion of an advanced rope capability in a full-mission-profile exercise.

Props can also be used to enhance any scenario. A facilitator can make great use of items such as high-pressure air lines that leak on command, simulating a release of natural gas or other substances. A strobe light can be used to create distractions and simulate electrical malfunctions, and you can add aural authenticity to an industrial area by piping in high-pitched noises to simulate the din of machinery. Fake panel boxes are useful for training personnel in lockout, tag-out procedures, and pipes with various sorts of valves can similarly be used for bleed operations.

The training for any organization should be conducted on four levels: Awareness, Operations, Technician, and instructor. As mentioned earlier, in deference to NFPA 1670, each organization must decide whether it will adhere to a standard-specific curriculum. For a municipal department contemplating a Technician level of ser-

A lockout, tag-out wall at the FDNY training facility.

vice, this is undoubtedly the best path to follow. Industrial brigades, however, are more apt to provide training limited to the scope of the hazards present at the worksite.

Technician-level personnel will make up the key components of your response team. Both their level of skill and command responsibility set them apart. At first glance, the requirements for the Technician level of training may seem to be little more than those for the Operations level. Nothing could be further from the truth, as trainers and technicians can attest. Not only is command and con-

58

trol of the team left to Technician-level of personnel, but also all of the complex and long-distance entries that fall outside the auspices of the Operations level. This translates into much more training for advanced entries into tanks, vaults, and other difficult areas. There is a great difference in making an entry into a space where the victim is easily retrievable and one where the ingress and extrication distances may involve hundreds of feet, elevation changes, and an untold number of barriers. Rigging an overhead ladder derrick to remove a visible victim from a tank car is one challenge, entering to retrieve him from the farthest recesses is another. Additionally, Technician-level personnel are responsible for all of the planning and strategic analysis of an operation. Don't let the wording of NFPA 1670 fool you. Look at the appendices and realize that everything that Operations personnel cannot do, Technicians must.

Recurrent training is essential for all Technician-level personnel. Aside from in-station or in-plant drills each shift, the team should be required to train at least once per quarter, preferably more, for an entire day. These quarterly drills should include classroom instruction, workshop sessions, exercises with a full-mission profile, and any update information needed to keep personnel at their level of capability. At no time should you allow personnel, Technician or otherwise, to train to the minimum standards. Keep them all above par.

The requirements for instructors aren't implicit in NFPA 1670, but instructors should be drawn from your technician base, usually directly from the team. Each organization has its own requirements for becoming an instructor. Suffice it to say that there is no substitute for knowledge gained from a mature individual with plenty of firsthand experience. Remember that instructors represent the future of your program. If knowledge isn't passed on, it's wasted. By the same token, instructors who aren't actively involved with the team in the operational theater risk losing touch with advances in the field, and nothing will ruin your program more quickly or create more liability concerns than relying on poor instructors who provide out-of-date or erroneous information.

If you're going to provide the best level of service to the community, you must provide front-line medical care. This means having an ALS specialist on the team who can work in a confined-space environment. Experienced rescue technicians know full well, however, that you can't just put any paramedic or physician into the hole and expect him to administer good medical care to a victim. Confined-space environments are simply too hostile. Compound that basic unfamiliarity and disorientation with a lack of team training, and you'll be setting yourself up to injure a paramedic or physician. Technical-rescue medics should be experienced physicians or paramedics with a lot of street time who are then recruited to join a technical rescue team. Physicians with a strong background in emergency care, particularly in the emergency room, often make the best candidates, since they're accustomed to working under pressure and are typically inured to getting dirty. The medical implications of confined-space rescue, of course, include triage, unique patient-packaging schemes, lengthy extrication times, ventilation, amputation, and crush death syndrome, and these are concerns best left to competent medical specialists.

Technical-rescue medics and physicians must maintain all ALS requirements and licensure; must meet Technician-level requirements; must be active members of the team and receive recurrent training; and must meet any special training requirements outlined by the medical director of the organization.

CHAPTER QUESTIONS

1. Training that is realistic will present a certain amount of _____.

2. The objective above and beyond the ostensible purpose of any exercise should be the _____.

3. Where should any basic course begin?

4. What are the four key purposes of practical training conducted with a nonmission profile?

5. Why might an instructor coach a two-member team into getting through a maze as quickly as possible?

6. What is an in-and-out drill?

7. What is an exercise with a full-mission profile?

8. What is the difference between a managed exercise and a facilitated exercise?

9. The training for any organization should be conducted on what four levels?

10. Aside from in-station or in-plant drills, how often should the team should be required to train, at minimum?

CHAPTER FIVE

EQUIPMENT

Requirements for products used in hazardous locations have been in existence in the United States for at least eighty years. Although it is difficult to say who wrote the first of them, the compilers of the National Electrical Code (NEC) first addressed the installation of equipment for use in hazardous locations in 1920. In 1930, Underwriters Laboratories (UL) published its Standard 698, *Standard for Industrial Control Equipment for Use in Hazardous Locations.*

Within the United States, these requirements evolved around a single system of classification known as the Division system. Today, the Division system addresses the design, manufacture, installation, maintenance, and inspection of hazardous areas, as well as the equipment and wiring used in them. In Europe, independent development based on the International Electrotechnical Commission (IEC) resulted in the Zone system. Although the Division system is the accepted standard in North America, the vast majority of the world's hazardous locations are classified using the IEC Zone system. Both systems exist for essentially the same reason, however, which is to classify environments by the type of combustible present and whether it presents a continuous danger or not. A Division 1 area, for example, is one in which a combustible liquid, gas, or vapor may be present under normal circumstances. At the top of this system, flammable materials are divided into three classes. Flammable gases and vapors are included in Class I; combustible dusts such as grain and coal fall under Class II; and Class III consists of ignitable fibers, as are commonly produced in cotton mills. At the bottom of the hierarchy are groups of chemicals, designated by a letter of the alphabet. The materials in any particular group are so designated because they share similar properties of combustion, such as gasoline, acetone, naphtha, fuel oil, and methane, which fall under Group D.

Using such a system, a hazardous atmosphere can be routinely and adequately defined. A spray-painting operation using acetone, for example, would be classified as a Class I, Division 1, Group D environment.

As a result of pressures from the global market, it became critical for the United States to examine the Zone system more closely to find a way to integrate it into the Division code. In 1995, the United States completed this review and adopted the IEC Zone system into the 1996 edition of the NEC.

Why is all of this important to the confined-space technician? Obviously, you must be able to speak the language regarding the specification and purchasing of equipment. If you are a vendor, you need to be able to compete globally, selling equipment that has worldwide application.

Any device used in a confined space must be safe. Electrical devices must be constructed in a manner so as not to present a source of ignition in a combustible atmosphere. Several engineering, insurance, and safety organizations have standardized test methods, established definitions, and developed codes for testing electrical devices used in hazardous locations. The NFPA has created minimum standards in its National Electrical Code, published every three years. The code specifies what safeguards may be used in hazardous atmospheres, and it offers information on various materials that contribute to atmospheric combustibility. For the purposes of the NEC, an atmosphere is considered hazardous if the concentration of any combustible in it is within the flammable range, if a source of ignition may be present, and if the resulting fire could extend.

There are three methods of construction that can prevent potential sources of ignition from escaping and igniting a flammable environment. An ignition source that is encased in a rigid container is considered explosionproof. If an arc is generated, the ensuing explosion is contained within the enclosure. Devices in which the problematic components are encased in a solid insulating material are also considered to be intrinsically safe. In a purged device, the third type, a steady stream of a noncombustible gas is used to act as a buffer, keeping the flammable atmosphere away from the ignition source.

Additionally, there are a number of lesser construction methods, subordinate to explosionproof and intrinsically safe devices, used to ensure electrical safety. These include nonhazardous circuits, nonhazardous components, nonsparking apparatuses, hermetically sealed components, flameproof components, dust-tight components, and the like.

Certification means that a given device has been deemed explosionproof, intrinsically safe, or purged for an atmosphere of a particular class, division, and group. The device is not certified for use in atmospheres other than those indicated. All certified devices must be marked to show class, division, and group. Any manufacturer that wants to have an electrical device certified must submit a prototype to a laboratory for testing. If the unit passes, it is certified as submitted; however, the manufacturer also agrees to allow the testing laboratory to check the production line at any time, as well as any marketed units. Furthermore, if the product is updated or modified, the manufacturer must contact the laboratory for a new round of testing and certification. Testing is done by organizations such as Underwriters Laboratory (UL) and Factory Mutual (FM). NFPA does not do certification testing.

Some organizations test and certify instruments for environments that differ from the NEC definition of a hazardous atmosphere. The Mine Safety and Health Administration (MSHA), for example, tests instruments only for use in methane-rich atmospheres containing coal dust.

Certifications issued by laboratories elsewhere in the world, including Europe, aren't honored in the United States without separate U.S. approval. The exception to this rule is Canada. Any equipment approved by the Canadian Standards Association (CSA) must carry CSA on its label, with NRTL appearing below. This indicates that OSHA has approved the equipment for use in the United States, based on adherence to test methods specified by UL and ANSI.

Obviously, the test protocols for Division 1 certification are more stringent than those for Division 2. Thus, a device approved for Division 1 is also permissible for Division 2, but not vice versa. For most response work, technicians should choose devices approved for Class I areas (liquids, gases, and vapors), Division 1 (areas of ignitable concentrations), and Groups A, B, C, or D whenever possible. At minimum, an instrument should be approved for use in Division 2 locations.

There are so many groups, classes, and divisions that it is nearly impossible to certify an all-inclusive instrument. Therefore, you should select a given device based on the chemicals and conditions that you expect to encounter, and purchase certified instruments for any high-risk or specialty spaces to which you might respond.

Every piece of equipment that a technician uses is, in effect, personal protective gear, since it will keep him alive and operating in a hostile environment. When dealing with an environment containing haz mats, any protective ensemble will have to be selected based on the chemical data. If you and your team intend to operate in chemical environments, you'd better have adequate research capabilities to rely on.

A low-profile helmet is essential for both head protection and ease of movement. Eye protection in the form of goggles may be incorporated into the design. Face shields aren't safe, since they let in debris from underneath. Fire service helmets are too bulky and won't allow freedom of movement in areas that are truly confined. Each equipment cache should have a supply of low-profile helmets for those really tight spaces.

Besides hypoxia and asphyxiation, the greatest threat to the confined-space rescuer is the flash fire. Naturally, no one should be knowingly enter an explosive or flammable atmosphere, but circumstances can change rapidly in a confined space and you may not be able to control the environment as much as you wish. Whatever the reason, protection of your head, ears, and face is critical. Fire-resistant hoods also offer some degree of protection against extremes of temperature. Flash protection for your body is equally vital. Long sleeves can merge with the gloves to provide protection for the entire arm and hand. The first level of protection, of course, is fire service bunker gear, but this is inappropriate for entry personnel. Those working outside the space may wear it, though even there it is less than ideal. Entry personnel should be in fire-resistant jumpsuits, some varieties of which are actually made out of the same material as turnout gear. Ideally, the jumpsuit should be reinforced at the prime impact points, such as the elbows, knees, and buttocks. Even so, technicians should wear knee and elbow pads. Specially fabricated coveralls are available for additional protection against certain chemicals.

A simple pair of earplugs for use around machinery or inside spaces with a high level of noise is necessary. In most instances, the breathing apparatus mask will act as eye protection, but when working outside of the space, goggles or approved safety glasses are also a must.

Two types of gloves are required, an inner glove for manual dexterity and an outer work glove. A high-quality leather work glove, not a firefighting glove, will offer good protection against mechanical injury; fire-resistant flight gloves will afford the wearer a fine sense of touch. Although the latter will tear easily when used as a work glove, the leather gloves can be taken off when greater manual dexterity is required.

Completing the body, a quality leather work boot that provides good ankle protection and steel toes is essential. Whoever the manufacturer is, make sure that the boot is comfortable, that it provides good ankle support, that it has good traction, and that it has steel toes.

All members of the entry team need to wear a Class III, full-body harness. A harness makes it possible to enter and exit a hole by way of a rope system. During entries and retrievals in the vertical mode,

An entry-team member with all necessary PPE and equipment, minus victim SABA.

each technician must be attached to a system that can arrest him if he falls, as well as raise and lower him. Once he is inside the space, he may disconnect this system and leave it at the point of entry.

Lights are also necessary, and they must be appropriate for the environment that you're entering. All entries must be made with either SCBA or SABA, regardless of the results of atmospheric monitoring. Always remember that SABA is made for confined spaces, SCBA is not. Still, each team member, or entry team at the very least, needs to carry a personal monitor. At minimum, the monitor should be capable of monitoring flammability, toxicity, and oxygen levels. Litmus paper is also nice to have and often necessary.

Entry technicians must be in constant contact with the outside, and a hardwired, hands-free system is preferred. A rescuer should also be equipped with a personal alert device that will activate automatically in the event that he stops moving. Only the first team needs to run a tag line with them into the hole. A tag line is a Hansel and Gretel device. You take it to the farthest point of penetration and anchor it, or you incorporate it into your SABA or communications umbilical. In truth, once you make a few turns, bends, and drops in a confined space, a tag line isn't going to help you to get out again.

Every rescuer should carry a couple of carabiners, a 24-foot section of two-inch tubular webbing for a hasty hitch, and a collapsible cervical collar for quick access to the patient. You should also carry some form of respiratory protection to place on the patient when you gain access to him.

There are basically four types of respiratory protection available on the market. Similar to a gas mask, an air-purifying respirator (APR) is either a half- or full-face mask with filters on either side. The filters come in different varieties and may be changed, depending on the contaminant encountered. These devices may not offer an adequate level of protection and should never be used in an IDLH environment (imminently dangerous to life and health). If the atmosphere that a technician is going to work in is low in oxygen, an APR will have absolutely no value whatsoever. Also, this type of respirator can only provide protection if the filter is compatible with the type of atmosphere in which the user is going to work. He must first determine what toxic gas is present, and in what concentration, before he can verify whether the mask will protect him or not. Although they have their place in industry, APRs should never be used for rescue operations involving IDLH environments.

The standard fire service breathing system is SCBA, or self-contained breathing apparatus, with air tanks of thirty-, forty-five-, or sixty-minute capacity. The range of tank pressures runs from 2,200 psi to 4,500 psi. Each system includes a backpack, bottle, and mask. On newer versions, the regulator is mounted on the mask. SCBA was designed for structural firefighting, and the tanks were never intended or designed to be removed from a firefighter's back while in use. These systems are large, they have a limited air supply, and they are ungainly for use in a confined-space environment. Some newer versions do have in-line, supplied-air capabilities, and these are suitable for certain confined spaces.

Rebreathers offer breathing times of up to four hours. These devices have a small liquid oxygen container and a scrubber that allows the wearer to rebreathe whatever air he uses. The scrubber removes carbon dioxide, and fresh oxygen is provided by the liquid oxygen. These devices are excellent for use in mines, tunnels, and other locations where travel over great distances is necessary, but they have the same limitations as regular SCBA for true confined spaces. They are large, and they aren't meant to be removed from the back. Because a rebreather operates on a closed system, it is even bulkier than SCBA.

The ideal system for confined spaces is SABA, or supplied-air breathing apparatus, since it was designed specifically for these types of operations. It consists of six components, namely the bottles, supply hose, first-stage regulator, manifold, mask-mounted regulator, and the face piece. The supply side of the system provides unlimited air from an air-cart system, breather box, or a cascade or mobile unit. A medium-pressure hose runs from the

The ideal breathing system for confined spaces is SABA.

supply side to the user. Polyvinyl chloride hose is viable at temperatures from 32°F to 120°F, neoprene hose is viable from -25°F to 212°F, and nylon hose is viable from -25°F to 120°F, so choose a type that is compatible with the environment that you intend to enter or work in. SABA provides unlimited quantities of air from the supply side and a backup bypass bottle for emergencies. The bypass bottle may contain from five to fifteen minutes' worth of air, and the regulator may be belt- or mask-mounted.

There is an ongoing debate about using SCBA rather than SABA in confined spaces, and arguments abound concerning the merits of one system over the other. One common argument centers around the fallibility of any mechanical device, with the implication that complex systems are correspondingly more prone to failure. Still, SABA comes equipped with a secondary air supply, affording the user some escape time in the event of failure. Standard SCBA lacks a secondary air system. Also, SABA needn't necessarily be any more prone to failure than any other type of breathing apparatus, given that any respiratory protection is only as good as the maintenance program and the competence of those who use it.

The average SCBA bottle weighs from fourteen to twenty pounds, depending on the materials of construction. Once you remove the tank for any reason, it becomes a weight just begging to be released. If it falls away, it will pull off your mask, and in a hostile environment, that could at once prove deadly. Many techniques have been developed and are still being taught, by certain instructors, for removing the bottle in a confined space. The prusik method of attaching the bottle to the main haul line precludes dropping it during entry or exit through a cramped portal. The problem with this method is that you'll have to take the bottle off the rope again to put it back on, and in a confined space, this may be a daunting task. God forbid that you should have to get out through that opening in a hurry! Some instructors advocate the horizontal or vertical buddy-pass method, in which one team member passes the SCBA tank to another member already in the hole. In the real world, however, there never is someone conveniently waiting inside the hole to assist the first responder.

NFPA 1670, *Standard on Operations and Training for Technical Rescue Incidents,* indicates that equipment shall be used in accordance with the manufacturers recommendations. To my knowledge, no manufacturer indicates that a standard SCBA is designed to be taken off the user's back during normal operations.

The requirements of respiratory systems are outlined in OSHA standard CFR 29 1910.134, *Respiratory Protection.* This standard provides baseline information on the use and selection of respiratory equipment, but it falls short of specifically requiring the use of SABA for confined spaces. It does state in paragraph (b)(11) that,

"The respirator furnished shall provide adequate respiratory protection against the hazard for which it is designed in accordance with standards established by competent authorities." It can be argued that SCBA, which only designed to meet the demands of structural firefighting, does not meet the requirements of this standard. Admittedly, this is only an interpretation.

NIOSH does not write standards. This organization merely approves equipment that meets its guidelines. According to its *Certified Equipment List as of December 1991,* a system that provides air by way of an air line is approved for entry into oxygen-deficient atmospheres, whereas a self-contained air supply is approved for escape only.

SCBA clearly does have its place and uses in confined-space operations, but mostly by response personnel outside the space, acting in a support role. If the space is big enough to walk into, and if the apparatus doesn't have to come off the wearer's back, then it does have some limited applications inside the theater of operations as well. The in-line attachment available on some models provides the equivalent of a thirty-minute bypass bottle. This is a great system, but it still has limitations in tight spaces. If the apparatus has to come off your back for any reason while moving around the space, then you are using the wrong tool, and it could prove deadly.

Despite all the arguments, the final determinant is that SABA was designed for confined spaces and SCBA was not. Budgetary concerns are no excuse for failure to obtain the right equipment to do the right job. Everything has a cost, and the cost of SABA is far less than that of a team member's life.

Along the same lines, risk/benefit issues surround the practice of buddy breathing as well. Of course, if everything else fails, a rescuer is going to do whatever it takes to get out of the space alive. As in all other technical endeavors, proper procedures and equipment are required for buddy breathing. First, you need to ascertain that you have a NIOSH-approved buddy-breathing port on your apparatus. Second, you need to train, train, train with it, in every possible configuration—in the dark, with gloves on, lying on your back, and in every other mode that you can conceive. All of the breathing apparatus used by the team must be compatible. Typically, the breathing apparatus of different manufacturers are not compatible, so beware.

As mentioned above, you must able to monitor the space for atmospheric hazards. These include, but aren't limited to, oxygen content, toxicity, and flammability. Reliable detection devices are critical toward ensuring our safety when operating in confined spaces. It's virtually common knowledge that the miners of yesterday took canaries with them into the mines, and that when the canaries went feet-up, it was a clear indication that the air was either toxic or deficient in oxygen. Perhaps that technique isn't so outdated, for anyone who saw pictures taken in the aftermath of the sarin attack in the Tokyo subway may have noticed that response personnel were carrying cages with small birds in them.

Detection technology has come a long way, of course, since Sir Humphrey Davy (1778-1829) invented the Davy safety lamp. The Wheatstone bridge combustible gas indicator, introduced in the 1930s, was another significant step forward. Today, the pace of technological advancement is marching so fast that it would be futile to mention all of the features and options available on modern monitors. These devices aren't cheap, averaging about $2,000 and requiring significant maintenance and care. You might as well accede to the notion that, unless your organization is blessed with unlimited funding, you're going to have to live with whatever technology you purchase for some time.

When purchasing a monitor, there a number of points to consider. The unit should be durable, and if you anticipate using it in a wet environment, it should be water resistant, if not waterproof. It should be a handheld, portable device, capable of detecting oxygen levels, flammability, and toxicity. It should have both a loud audible alarm and a visual indicator. Naturally, certification regarding the Class, Division, and Group of environment you're working in is vital. The monitor must have either a mechanical or a hand-aspirated pump for drawing samples, and the display should be easy to read, with the detection parameters clearly marked. It must also be easy to calibrate in the field and capable of clean-air setups. It should be RF shielded. A users manual that is easily comprehensible is absolutely necessary, and it's always best if a manufacturer's representative will provide training for team members as part of the purchase price. A maintenance program and aftermarket support are also necessary.

Monitors come in a variety of sizes and shapes, but all share some common components. The device itself is housed in a hard case that holds the electronics and sensors. The readout may be either digital or analog. The controls typically include an on/off switch, zeroing adjustment, clean-air setup, range selector, span adjustment, battery check, backlight, and an alarm. All of these components are controlled by way of a keypad, knobs, switches, or buttons. Sample air is drawn into the instrument by either a vacuum pump or manually, using an aspirating bulb. Often the terminal end of the sample line has a filter to keep out liquids and other foreign matter. The unit can also analyze air that passively comes in contact with its sensors. A battery provides power to the device, with a charger available.

There are several methods by which a monitor can detect atmospheric hazards. The electrocatalytic method is based on the explosive range of a given substance. By design, of course, a gas monitor should emit a warning before the concentration of a given flammable gas reaches the lower explosive limit. By providing a heated catalyst to known combustibles, an electrocatalytic gas sensor can measure the amount of combustible present. When a flammable substance contacts the catalyst, combustion occurs. The resulting release of heat raises the temperature, as well as the resistance, of a coil of fine platinum wire, which also acts as the temperature sensor. To guard against variations in ambient pressure and temperature, a compensator device, insensitive to combustible gas, is incorporated into the system. This provides an electrical signal indicating the amount of gas in the sample.

A semiconductor detector employs a metal oxide sensor that can detect low concentrations of certain combustible gases. As heater coils maintain a constant temperature, gas is chemically absorbed by the surface of the sensor, affecting its electrical resistance. These minute changes are the basis for electronically calculating the concentration of the gas.

In an electrochemical detector, gas is admitted into the device through a semipermeable membrane, where it reacts with an electrolyte solution. The resulting flow of ions indicates either the partial pressure of oxygen or the concentration of other gases. By the inherent nature of their design, electrochemical cells are stable devices, capable of operating for as much as a year or two without maintenance. Consequently, they are useful for monitoring toxic gases or oxygen deficiency over time.

Infrared detectors measure the presence of certain gases by quantifying the absorption of infrared radiation passing through the gas. Some gases are markedly more opaque to infrared radiation than they are to visible light. Thus, a sensor that can measure the diminishment of radiation from a known source can be used to calculate the concentration of the gas.

Photo-ionization detectors, on the other hand, use ultraviolet light to excite gas molecules, temporarily stripping them of an electron and creating an ion with a positive charge. The electric current produced in this process can be measured to find the amount of gas present. Once the source of ultraviolet light is curtailed, the electrons return, de-ionizing the molecules. Thus, there is no contamination inherent to the process, as there is with reactive cell or sensor devices. One drawback of this sort of monitor, however, is that it may be affected by high humidity. It will also only detect volatile organic compounds (VOCs) within the ionization potential of the bulb.

Familiar to any high school chemistry student, litmus paper is used to measure pH, the relative acidity or alkalinity of a substance. Since acidic environments can destroy the sensors of expensive monitoring equipment, it's best to determine the overall

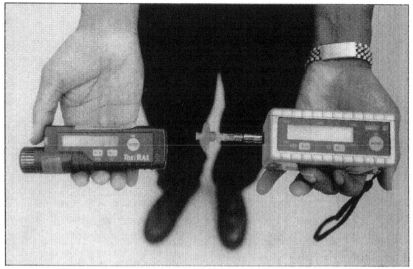

Two types of photo-ionization devices.

pH of the atmosphere before exposing your monitor to it. Place a piece of litmus paper on the end of a stick or some other probe. When the litmus paper changes color, compare it against the scale provided on the side of the container.

Similarly, colormetric tubes contain a catalyst that will change color when exposed to certain chemicals or classes of chemicals. They have been used by haz mat teams for years, but they have some serious limitations. They provide only a snapshot view, for one thing, and they are incapable of continuously monitoring the air or analyzing the quality of it. They are slow to produce a reading, requiring minutes rather than seconds, yet their accuracy range is only within 25 to 35 percent. Colormetric tubes are expensive and, since they expire, any organization that relies on them must maintain a large supply.

Ventilation has been used in the fire service and by rescue teams for decades, and the concept of it is certainly anything but new. The purpose of ventilation in a confined spaces is the same as it is in a structure: to remove atmospheric contamination and replace it with clean, fresh air. Toward this end, any equipment that you purchase should be lightweight and portable, with a minimum capacity of 3,000 cubic feet per minute. It should be compatible with remote ductwork, either rigid or flexible. The flow rate at the end of the duct should be a minimum of 200 linear feet per minute. Depending on the environment in which you intend to use it, the equipment should also be intrinsically safe or explosionproof. Accessories are available for fitting ventilation equipment to certain types of spaces, such as manholes and tank trucks. The power source of the device, whether electrical, gasoline, or diesel, is a consideration.

Two types of fans are used for ventilation. A centrifugal-flow fan draws a stream of air parallel to the propeller shaft, then turns around 90 degrees and expels it. Centrifugal fans are efficient, but they cannot move large volumes of air. Several types of fan blades are available, including paddle-wheel, forward-curved, and backward-curved blades, each designed for a specific task. Paddle-wheel blades are flat, used for medium volume and medium speed, high-pressure applications. This type of configuration is very good for moving air containing particulates. Forward-curved blades

A typical axial-flow fan.

consist of many narrow, curved blades set in a shroud ring. These aren't suitable for air containing particulate matter or other substances that can clog the blades. Backward-curved blades can be flat or curved, and they can form part of the rotor. These are used for large volumes and high-speed applications.

Axial-flow fans, the second type, draw air inward and discharge it back along the propeller shaft. Often found on fire apparatus, some axial-flow fans on the market are notable for their new and unconventional designs, bearing no resemblance to the typical fire service variety. A simple axial-flow propeller is a two- or three-blade propeller mounted on a shaft. This type of fan is used for moving large volumes of air at low velocities. It does not produce enough flow to force air through ductwork. A tube axial-flow fan is characterized by a propeller mounted within a cylinder. Depending on the diameter of the fan and the power of the motor, this type can move large volumes of air at medium pressures. A vane axial-flow fan is similar, but the vanes direct the air in a straight line. This type is relatively small and light, given the amount of air that it can move.

An air ejector may be another option. These devices use the venturi effect to move air from a given space. Air or steam is blown through a tube, creating a low-pressure area. Large quantities of air are thus drawn into the tube. Ejectors that rely on air can be used for either supply or exhaust, while those that rely on steam can only be used for exhaust. In general, ejectors are lightweight and portable, and they can easily be connected to ductwork. They may be used in explosive atmospheres or wherever matter may clog a fan. One drawback, however, is that they require large amounts of air or steam to operate, and they create static electricity, requiring them to be bonded and grounded.

Accessories are a must, so when purchasing the right fan, make sure that you get the right attachments to go with it. Saddle vents are attachments that fit on the terminal end of flexible ducts, allowing you to place the vent inside a manhole. A saddle vent will still allow for access to and from the space without removing the ventilation, and it doesn't reduce the end flow from the duct. The duct-work itself can be either flexible or rigid. Be sure to get the correct connections for fitting ductwork to the fan.

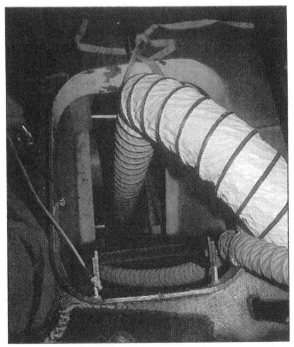

Flexible ductwork.

Long before safe and reliable communications systems were available, confined-space rescuers relied on shouting and a vocabulary of rope pulls. The old OATH method was perhaps the one most commonly taught. By this method, one tug meant *Okay,* two meant *Advance,* three meant *Take up,* and four meant *Help.* Anyone who has ever tried communicating by rope knows how difficult it is to decipher the number of pulls coming through the line, especially when the team is moving and the rope has passed around a couple of corners or has gone vertical.

Instantaneous communication between the rescue team and personnel on the outside is critical to both the safety of the team and the success of the mission at hand. Vital information regarding atmospheric monitoring, air supply, patient status, ventilation problems, requests for additional resources, and all else must freely pass between those inside and those outside the hole. Naturally, the system you choose must be intrinsically safe or explosionproof, as appropriate to the environment. It must also be able to withstand the rigors of the job, ranging from moisture and chemicals to rough handling. Further, it must be viable when surrounded by dense structural materials, such as steel and concrete. Electrically powered communications equipment falls into two basic categories, wireless and hardwired. The highest rating of safety for intrinsically safe communications equipment is Class I, II, and III, Division 1 and 2, Groups A, B, C, D, E, F, and G.

The advantages of a wireless system are obvious. The freedom of movement that it affords is doubly precious, given that the rescuer may already be dragging a SABA line with him through tight quarters. A wireless system, however, is subject to dead spots and intermittent communications, and it is not a hands-free system, since it requires the user to push a mike button to talk. Voice-only accessories require fine adjustments and can lock open the system in a noisy environment. Transmitting by radio may affect other instruments, and the message itself may be garbled at the outset simply by having to communicate through a face mask. A radio channel can be defined as an inherently nonprivate means of communication that can be interrupted by other transmissions. Users on the same frequency can all too easily step on each others' calls. Radios used in confined spaces are subject to considerable damage, meaning high repair costs, and intrinsically safe radios have a limited level of approval.

The principal disadvantage of any hardwired communications system is the wire itself. Those who engage in activities or operations in which the user may roam free would never opt for a hardwired system over a wireless one. This is not true of confined spaces, however, where hardwired systems can be used effectively and provide a high degree of reliability. Any hardwired system can be mated with the breathing apparatus supply line and tag lines simply by encasing them all in a section of two-inch tubular webbing and binding them with wire ties, thereby forming a single umbilical. A hardwired system can provide a clear, continuous, hands-free, dedicated link between the parties. As long as the line is properly shielded, their communications won't be affected by electrical interference. The maintenance and repair costs of a hardwired system are relatively low, and a high level of safety approval is available. If you have access to military surplus facilities in your area, you may find some older-style sound-powered phones. If your team cannot afford a modern system, these are a great low-cost option.

Communications equipment designed for diving operations or high-angle operations is sometimes used for confined spaces. Such equipment employs communications wires embedded in kernmantle rope. Since the OSHA standard no longer requires a tag line, this may or may not be a consideration to the user. Although offering the same benefits as a hard wire, the life of the rope is limited, and the replacement costs are an ongoing concern. Life safety rope that is shock loaded must be removed from service, and you will be removing your communications line as well. Moreover, the wires can't be inspected for damage, and diving equipment doesn't typically come with the requisite safety approval rating.

Retrieval systems are used by industrial workers who enter confined spaces, and also by rescue teams. These devices allow the support team to insert a worker or rescuer vertically into a confined space and to haul him out again. They are also used to raise victims. A fall-arresting system may be incorporated into a retrieval system, either as a factory feature or as an aftermarket modification. In specifying and making purchases, remember that, to meet the needs of different types of locations, your team will likely need to purchase more than one type of device. For the purposes of this text, we will consider retrieval and fall-arresting systems together, since you will need both throughout an operation. A good system incorporates both into one piece of equipment.

A retrieval system may include a chest or full-body harness (Class III), wristlets, and some type of lifting device. In the va:t majority of incidents, you will need an overhead anchor point, to which you can affix both your haul and fall-arresting systems. In some instances, industrial sites have overhead anchors already in place over certain spaces. Even if there is a ladder leading vertically into the recesses of the space, all who enter should be placed on a fall-arresting or belay device in case they slip.

Every team should have at least one tripod for use at manholes and other similar openings that can be found in any response area. A number of manufacturers make them, both aluminum and steel. Most tripods have telescoping legs, enabling the device to be extended up to sixteen feet. They also collapse into a storage bag that houses not only the tripod, but all of the accessories as well. The overhead anchor point that a tripod affords allows for the connection of a rope system or a commercially available cable fall-arresting system. Many tripods come with a manual cable winch that may have an extension arm that provides greater leverage.

Tripods do have some significant drawbacks in certain environments. Perhaps most significantly, they require the user to pull the load directly in the center; otherwise, the three-legged device might tip over. This means that, if a change of direction is required in your haul system, you may need to anchor the tripod to the floor. Also, since tripods require an equal stance for all three legs, you may not be able to set one up in close quarters, against a wall, or on a curved surface.

Another option is a single-arm davit, a device with anchors at the bottom, supported by two horizontal legs. The arm may be pivoted, allowing it to swing anywhere from 45 to 180 degrees, left or right. Various accessories will allow you to use this type of device on tanks, ledges, and other hard-to-reach areas where a tripod would not be viable. Like the tripod, however, a single-arm davit will require some sort of anchoring if off-center hauling is to be done with a change of direction. Also, these devices are large and difficult to store on response apparatus.

Other overhead anchor systems may include ladder derricks constructed of fire service ladders; an aerial ladder or bucket truck; and industrial or marine anchors over certain openings and hatches.

An overhead ladder derrick.

If you have a highly proficient team that trains often and works well with equipment, some brand of prerigged mechanical advantage system will offer the most flexibility. Preconfigured systems set up to provide 3:1, 4:1, and 5:1 advantage are available. These also provide a rope friction system and are typically used in conjunction with a tripod or some other overhead attachment point. If, however, your team doesn't train that often or isn't proficient with rope systems, then a mechanical winch and wire rope system is best. These require little training and can be used by just about anyone.

All ropes, harnesses, and hardware that you purchase should comply with NFPA 1983, *Standard on Fire Service Life Safety Rope and System Components*. All of your rope should be either nylon kernmantle or polyester rescue rope. There is no longer any place for manila rope in rescue operations. It's your decision whether to use 1/2-inch or 5/8-inch rope. Our team typically uses 12.5 mm for basic entry and for the lowering systems that involve patients packaged in a transport device, as in a rescue from a tower, off the side of a ship, or in many workplace locations. It's preferable to have both static and dynamic rope in your inventory. The static line will be used for lowering and hauling systems. The dynamic rope will be used if your team needs to traverse, climb, or set up a dynamic belay. In many rescue scenarios, a member may need to climb so as to set up a viable anchor point.

By standard, all personnel who enter a space with a vertical drop of more than five feet must wear a full-body, Class III harness. These offer both front and rear connections, as well as good stability during entry and exit. A wide range of manufacturers make a variety of sizes and shapes. Padded versions are recommended.

It's also important to have a number of rappel harnesses (Class II) for use by support personnel. These harnesses are used to provide an anchor point for their wearers when working high above the ground. Anytime that you are working at a site that presents a potential fall hazard, regardless of the distance, every team member should be in a harness and attached to a safety tether. Class III harnesses work just as well for this, but they are more difficult to get into than a standard Class II harness. Again, make sure to get a good, padded harness.

Carabiners, sometimes called snap links, are trapezoidal devices of aluminum or steel, used to connect other pieces of a rope system together. Make sure that you purchase pin-locking steel or aluminum locking carabiners. There is absolutely no place for non-locking carabiners in the rescue environment. For ease in connecting systems together, large and extra-large sizes are recommended.

Aluminum carabiners, of course, are much lighter than steel; however, they are also softer than steel, and they tend to wear out faster when subjected to friction. Aluminum carabiners develop heat faster than steel, but they'll also dissipate it more quickly. Steel offers extra strength and abrasion resistance, and it will develop heat more slowly under friction, but it'll dissipate that heat

Attachment point on a rescue worker, showing Class III harness and sheathed air-line umbilical. Strain relief is provided by 9 mm cord and a carabiner. The communications line runs over his right shoulder. Note that the comm line and tag line have been left out of the umbilical in this setup. This may pose additional concerns in the space. A bypass bottle is slung at the rescuer's right side.

more slowly. On the plus side, stainless steel carabiners don't spark, and they aren't magnetic. Aluminum will oxidize, and it can create a spark as well. If you're operating in an environment where sparks might cause a problem, your choice between these metals should be obvious.

In your arsenal of equipment, you should have some firefighter carabiners for quick attachments. These are rather large devices that look like the old ladder belt connections. They are excellent for connecting patients to hasty hitches or whenever rapid connections need to be made for movement or control. They're easy to work with while wearing gloves, and they offer a wide opening when the gate is fully open.

A variety of single and double pulleys are needed to create haul systems. They are also used to create changes of direction in the haul rope. Ideally, pulleys should have a sealed ball bearing and be at least four inches in diameter. Make sure that you incorporate double pulleys with beckets in your inventory so that you can create the systems you want.

"Descent-control device" is a broad term used to describe a myriad of devices that use friction to control the rate of descent. The two most common descent-control devices are the figure 8 with ears, commonly called a rescue 8, and the six-bar brake bar rack. Like carabiners, these are available in both steel and aluminum. It might be necessary for personnel to use these devices for rappelling, but more often you will use them as braking systems and belay systems. It's far more common for rescue personnel to be lowered to and raised from their area of operation than it is for them to reach it by rappelling.

As their name implies, ascenders are used for moving up a rope. They're also used for connecting personnel to patient-packaging equipment, such as a stokes tender line when acting as a litter attendant. Mechanical varieties, referred to as hard ascenders, include the direct-loading cam and toothed varieties. Soft ascenders are those made from rope, as typified by prusik hitches. It has been my experience that hard ascenders should not be used as braking systems or rope grabs in either hauling or lowering systems. Tandem prusiks are preferable for these applications. Still, you should have both hard and soft ascenders in your cache.

Edge rollers are devices used whenever rope passes over a hard edge, protecting it from cutting and abrasion. They come is several types. Sections of carpet, hose, or some other soft material can also be used to pad rough edges. Often this type of edge protection can be molded to unique surfaces where commercially manufactured edge protection won't fit or won't work.

Webbing is also an important part of rope systems, the most common being one-inch and two-inch tubular. The two-inch version is preferable, since it adds extra strength and width to your rigging

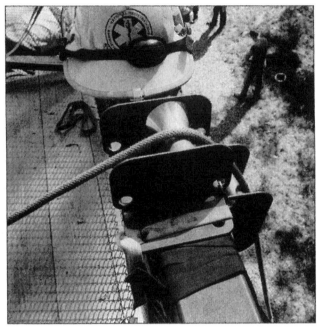

A commercial edge roller.

systems. Webbing is used to create a variety of connection points, on which you can build other systems. Webbing can be used as a loop around an anchor point, allowing you to connect a carabiner and a rope to it. It can also be used to create emergency egress seats. Its most useful application in confined spaces, however, is as a hasty hitch. This is a 24-inch section tied with a water knot to form a loop. This loop is used as a quick attachment point on a victim, allowing you to haul him out of the space. Webbing can also be used to create mariners hitches, which are load-releasing systems.

Small-diameter kernmantle cordage of 6, 7, or 8 mm is used for any number of utility purposes. Soft ascenders are typically made from this sort of utility cordage. It can also be used to create the hokie hitch, which, like the mariners hitch, is a load-releasing system.

During confined-space operations, you will face a variety of hazards that require identification and control. The purpose of lockout, tag-out and blank-out equipment is to eliminate a source of energy that might injure or kill a rescuer. These energy sources may be electrical, mechanical, hydraulic, pneumatic, thermal, or the like.

Lockout is a process by which an energy source—whether electrical, hydraulic, pneumatic, chemical, thermal, or otherwise—is shut down and rendered safe. Any energy-isolating device is capable of being locked out if it has a hasp or other means of attachment to which a lock can be attached, or if it has a locking mechanism built into it. The simplest example of an energy-isolating device that can be locked out is a manual electrical control box. In this device, you can pull down the main arm, thereby deenergizing the related circuits and machinery. Additional protection may be granted by fixing a lock that maintains the arm in the off position. Lockout devices may also incorporate padlocks, combination locks, chains, and the like.

Once a device has been locked out, it is also important to ensure that all personnel understand that the device should not be reenergized. Tagging out refers to the act of placing a tag on the device as

The common process of locking out and tagging a simple in-line valave could save someone's life.

89

a warning. Most such tags have bright colors and a warning, as well as a signature or other identification to show accountability.

Every member of the entry team should have his own lock and key. Once a lock is placed on an energy-isolating device, it should only be removed by that same individual. This will limit the chances that someone else might energize a dormant system while personnel are still working around it.

All lockout devices and tag-out devices must meet certain standards. When you set up your lockout, tag-out system, make sure that you place into service devices that will be compatible with the equipment for which they are intended. All lockout and tag-out devices need to be dedicated to the task of controlling energy sources and must not be used for any other purpose. Naturally, lockout devices must be able to withstand the environments in which you place them, and their tags should not deteriorate or otherwise become unreadable. All lockout and tag-out devices that you purchase must be standardized to your system, meaning that they should all be the same color, size, and shape. The print and format of your tag-out devices should also be standardized. Lockout devices need to be substantial enough so that no one can inadvertently remove them. Similarly, the attachment used on a tag-out device should be nonreusable, with a minimum breaking strength of no less than 50 lbs. In design, they must be at least equivalent to a one-piece, all-environment nylon cable tie. On the tag, there should be a space where the user can be identified, and all need to have a clearly stated warning, such as *Do Not Start, Do Not Open, Do Not Close, Do Not Operate,* or the like.

In some instances, as when dealing with valves, a lock may not provide sufficient insurance against someone reopening the system. In such a case, you should use a blank, which is nothing more than a solid piece of metal or plastic, sized and shaped to fit across the section of pipe you wish to seal off. Many pipe systems have connection points where you can slide in the blank, thereby preventing the chemical, steam, grain, or other product from flowing past. When ensuring that a valve stays shut, use a secure device appropriate to the valve that you are isolating. The same applies for electrical circuits.

Most confined spaces are dark, noisy places, requiring some form of lighting. Explosionproof cord lights are a great source of illumination, but they do require an outside power source, and the cord can be a hindrance. All members of the entry team should carry a handheld light. In some instances, these can be clipped to your entry gear or helmet. Chemical lights are nonreactive and produce no spark. You should get the high-intensity versions. These have a limited life span, so you'll need to take quite a few. Chemical lights also make great path markers, indicating the way to members entering the space behind you. There are also stick-on types that you can mount on the walls. Every member of the team should carry at least two chemical lights on entry. Rope lights, consisting of a long string of individual lights, are also available, but these offer very little benefit for the cost involved.

Of course, when you finally reach the victim, you'll need to have some method of removing him from the space. Exactly how you do that depends on the situation at hand. Naturally, the two most important factors are how much room you have to work in and how much time you have to get the person out. A victim in an environment in which the atmosphere has been compromised will likely need to be removed immediately for resuscitation. Another common scenario is one in which the victim has been injured, and thus needs packaging and some C-spine control. A third common situation is one in which the victim needs to be removed from a height. The transportation device, of course, must be lift-rated and capable of being used along the vertical plane. Your team should have transportation devices to meet all three of these basic scenarios.

In many instances, you will have to remove the victim immediately, by any means. Three common means of doing this are by means of a hasty hitch, wristlets, or a harness, either Class II or Class III. A hasty hitch consists of 24-inch pieces of two-inch tubular webbing, made into a loop by joining the ends in a water knot. This device can easily be carried in by the rescuer and placed around a victim found in any position. The hitch can then be attached to a haul system by means of a carabiner. This method, of course, offers no C-spine protection, and it should be used only in a last-ditch attempt to save someone's life. Wristlets can be purchased commercially in padded and unpadded varieties, or you can make them from two-inch tubular webbing. These devices simply attach to the wrists of the victim and then to the haul system. The victim is then hauled out. Be cautious when using these with burn victims, especially when there are burns on the arms or hands. To gain an appreciation of this method and a respect for the victim, every member of the rescue team should, during training, be hauled out of a space by wristlets, just once.

The harnesses that you would use for victims are the same types that rescuers wear. If you choose this option, remember that they are sometimes difficult to put on someone in tight spaces. When time is of the essence, you'd better be good at it and know your harness.

Finally, any jack of any trade will learn a few tricks along the way. Always be on the lookout for ideas germinated through experience, as unconventional as they may seem. As one small example of what I mean, during my career as a confined-space technician, I have found a skateboard to be an effective transportation tool. A skateboard is useful when you have to perform a long haul down a smooth surface, as you may encounter in culverts and some maritime environments. Placing the victim on a rescue board, then rolled along on the skateboard, will save your back.

CHAPTER QUESTIONS

1. In the United States, the system of classification for the design, manufacture, installation, maintenance, and inspection of hazardous areas, as well as the equipment and wiring used in them, is known as the _____.

2. The vast majority of the world's hazardous locations are classified according to the _____.

3. What are the three methods of construction that can prevent potential sources of ignition from escaping and igniting a flammable environment?

4. All certified devices must be marked to show _____, _____, and _____.

5. Does the NFPA do certification testing?

6. What does it mean when a given piece of equipment has CSA and NRTL on its label?

7. What four basic types of respiratory protection are available for confined-space operations?

8. Name the five different types of atmospheric monitors.

9. Ventilation equipment used for confined-space operations should have a minimum capacity of _____.

10. What was the OATH method of communication?

11. What are some of the disadvantages of a wireless system?

12. What two types of rescue rope are currently acceptable for confined-space operations?

13. By standard, all personnel who enter a space with a vertical drop of more than _____ must wear a full-body, Class III harness.

14. Once a lock has been placed on an energy-isolating device, who should remove it?

15. When should a hasty hitch be used?

CHAPTER SIX

COMMANDING THE INCIDENT

Much of the success of any operation is based on two specific factors. First and foremost, did the original personnel on the scene make the right decisions at the tactical level, and did they implement the right plan? Second only to these actions, how well were you able to command, control, and provide accountability during the operation?

Incident command isn't rocket science, and without a doubt, there are at least three or four schools of thought about how to accomplish it, what to call it, and how to define the assignments. Despite all the hype, hoopla, and contention over matters of detail, incident command is like anything else you do. To get good at it, you have to practice. If you are a firefighter, the most efficient way to prepare yourself and your organization for command of a confined-space operation is to use incident management every day, on every incident to which you respond. Responding to the common calls is where you're going to get your practice and follow the learning curve. If you aren't using incident management on a daily basis, you won't be able to use it effectively during a confined-space operation.

The other prime practice opportunity presents itself during training. Every training evolution should be guided by a command presence. This way, the members will get used to the terminology and begin to practice their tasks and responsibilities in the roles they were assigned, whether operations officer, extrication officer, air-supply officer, or otherwise.

The information that follows isn't written in stone. It merely represents one way of doing business. Use what is appropriate for your system, and don't be dismayed by the job descriptions or the IMS flow chart. You may, as other people have, look at that flow chart and find yourself wondering whether it's a rescue team that is required or an entire army. Despite any initial perceptions, you will find that the number of personnel needed for a significant confined-space operation is staggering, almost three times what is required for a structure fire of equivalent magnitude. An operable system of command and control is a necessity. With practice, you can set up a confined-space IMS as quickly as any other, building on it as the needs of the incident dictate.

RESCUE OPERATIONS OFFICER

A rescue operations officer is assigned at all technical rescue operations. This individual must be a Technician-level member of the rescue team. His general duty is to assume control of the rescue site, coordinating the sectors within it during the course of the operation.

Duties and Responsibilities

1. Reports to the incident commander.

2. Locates the most advantageous position from which to manage operations.

3. Assembles and disassembles incident sectors as necessary.

4. Manages the rescue site and coordinates the sectors that have been assigned to it.

5. Assigns and briefs technical-rescue personnel and support personnel in accordance with the incident management plan.

6. Directs the extrication sector, rescue equipment sector, medical sector, suppression sector, and the air-supply officer.

7. Ensures that all entry teams are briefed.

8. Coordinates efforts with the responsible party of the site.

9. Coordinates operations with the utility company representatives.

10. Ensures that all utilities and lockout, tag-out procedures are accomplished.

11. Determines the needs of the incident and requests additional resources.

12. Reports information about special activities and occurrences to the incident commander.

13. Ensures that the confined space is mapped.

EXTRICATION OFFICER

The extrication officer is responsible for the actual placement of teams into the technical-rescue environment, as well as removal of teams and victims. Reporting directly to the rescue operations officer, he must be a Technician-level member of the team.

Duties and Responsibilities

1. Reports to the rescue operations officer.

2. Directs the entry team officer, panel team officer, shoring team officer, rig master, and the sector officer.

3. Mobilizes teams and demobilizes them when their tasks have been completed.

4. Assesses team needs and communicates them to the rescue operations officer as necessary.

5. Ensures the quality and completion of any operation that will directly affect entry and exit from any technical-rescue environment.

ENTRY TEAM OFFICER

Another Technician-level member of the rescue team, the entry officer is also responsible for the placement of teams into the rescue environment. His job entails coordinating efforts with medical teams, digging teams, and packaging teams as necessary to gain access to and remove the victim. This person is also responsible for backup teams.

Duties and Responsibilities

1. Reports directly to the extrication officer.

2. Oversees the assembly of equipment and outfitting of any personnel who enter the technical-rescue environment.

3. Ensures that members have the proper protective equipment and tools for the task.

4. Ensures the readiness of backup teams for every entry team.

5. Integrates medical team personnel into entry teams as needed.

6. Ensures unimpeded communications with entry teams while they are in the rescue environment.

7. Logs all personnel in and out of the rescue environment and accounts for all entry personnel.

8. Forwards requests for resources to the extrication officer.

9. Ensures that members reach and disentangle the victim.

AIR-SUPPLY OFFICER

Predictably, the air-supply officer is responsible for ensuring that the proper masks, hoses, pumps, and other air-supply paraphernalia are in place for use during confined-space operations.

The air-supply officer must anticipate any resources needed to maintain a viable air supply to the rescuers and victim.

Duties and Responsibilities

1. Reports directly to the rescue operations officer.

2. Ensures that the requisite air-supply equipment is on site.

3. Ensures that any tools and equipment necessary for repairs to air-supply equipment are on site.

4. Ensures that backup systems are in place for any entry team that enters a confined space.

5. Supervises the hose teams and supply personnel for air carts or cascade systems.

6. Coordinates the movement of any air supplies needed at the site.

7. Ensures the refill of all empty systems.

8. Anticipates and requests any resources needed to maintain air-supply systems.

MEDICAL OFFICER

A technical-rescue physician or a senior technical-rescue paramedic is responsible for providing basic and advanced care to victims trapped in a technical environment. Additionally, this sector is responsible for providing cross-trained paramedics and physicians to enter the environment and treat patients, assisting in their removal as necessary.

Duties and Responsibilities

1. Reports directly to the rescue operations officer.

2. Ensures that the proper equipment for treatment and packaging is available for placement into the environment.

3. Assembles technical-rescue paramedic teams for entry into the environment at the direction of the extrication officer.

4. Directs technical-rescue paramedics and physicians.

5. Ensures coordination between the rescue team and EMS personnel so that there is a smooth transition of patient care once the victim is removed from the environment.

6. Ensures that the proper treatment protocols will be continued by the local medical control personnel.

RIG MASTER

The rig master is responsible for developing and rigging any rope-related systems needed for entry or removal of either team members or victims.

Duties and Responsibilities

1. Reports directly to the extrication officer.

2. Works in conjunction with the rescue equipment officer to convey the requisite equipment to the rescue site.

3. Directs all rigging teams in the development of rope systems on site.

4. Ensures the safety and sound engineering of any rope-related system for hauling or lowering.

5. Ensures the protection of all rope systems from mechanical or chemical damage.

6. Coordinates additional resource requests from the extrication officer.

RESCUE EQUIPMENT OFFICER

The essential function of this member of the technical rescue team is to act as a supply officer at the team truck or trailer, responding to requests for equipment from the rescue operations officer. He manages all equipment and support materials needed for the operation.

Duties and Responsibilities

1. Reports directly to the rescue operations officer.

2. Organizes and maintains the proper flow of equipment to and from the rescue site.

3. Identifies support requirements for planned operations.

4. Processes requests for technical-rescue equipment.

5. Maintains an inventory of all equipment used or expended during an operation.

6. Ensures the safety of all personnel engaged in shuttling equipment to and from the rescue site.

The officers listed above represent only a portion of the entire IMS hierarchy for a confined-space operation. Their duties and responsibilities should be supplemented by officers filling more conventional roles, and others, delegated as appropriate to the demands of the incident. The flow chart below depicts a command structure that can be used at both large and small incidents. As an incident expands, and as your need for resources and personnel grows, you can correspondingly expand the organization. Always remember, however, that you must call for manpower and equipment during the early stages of an operation. When you need something is not the time to begin looking for it. Always have at least two plans in mind, possibly more, in case your original options don't work out. Emergencies occur in the real world, and it is in that environment where physical law and Murphy's law both display the worst sides of their nature.

102

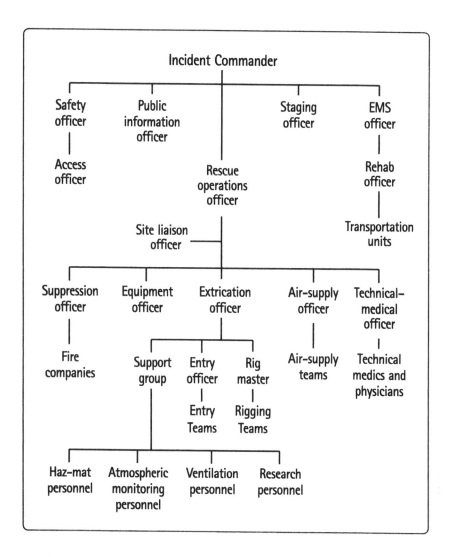

Checklists are vital for any complex undertaking. The following tactical work sheet is intended for both first-arriving personnel and the command staff. It provides all the aspects of an entry permit, plus it helps you gather all of the tactical data that you might need to make sound strategic decisions. The entry and medical checklists that follow are equally valuable for those particular sectors.

COMMAND TACTICAL WORK SHEET
CONFINED-SPACE RESCUE

___ **Primary Assessment**
    ___ Secure job foreman or responsible person
    ___ Secure confined-space entry permit
    ___ Determine location, number, and condition of victims
    ___ Determine whether rescue or recovery mode

___ **Secondary Assessment**
    ___ Type of confined space
    ___ Determine products in confined space
    ___ Determine hazards to rescuers
    ___ Assess need for additional personnel
    ___ Assess need for additional equipment

___ **Sectorize**
    ___ Safety
    ___ Hazardous materials
    ___ Extrication (technical rescue)
    ___ Operations (technical rescue)
    ___ EMS (treatment, transport)
    ___ Staging
    ___ PIO
    ___ Police liaison

___ **Rescue Operations**
    ___ Make general area safe
    ___ Make rescue area safe
        ___ Structural stability
        ___ Atmospheric monitoring
        ___ Ventilation
    ___ Lockout, tag-out, blank-out procedures
    ___ Entry personnel
    ___ Backup personnel (1:1)
    ___ Personal protective equipment
    ___ Air supply (SABA)
    ___ Communications and lighting equipment
    ___ Transportation devices for victim
    ___ Transfer to ALS treatment
    ___ Decontamination

___ **Termination**
    ___ Personnel accountability
    ___ Removal of equipment
    ___ Decontamination
    ___ Atmospheric monitoring
    ___ CISD
    ___ OSHA

CONFINED-SPACE ENTRY CHECKLIST

Location of Incident: _____ Purpose of Entry: _____

___ Secure responsible party
___ Secure confined-space entry permit/MSDS
___ Determine location, number, and condition of victims
___ Determine whether rescue or recovery mode

Type of confined space:
___ Tank   ___ Manhole   ___ Pipe   ___ Other

___ Assess need for additional personnel
___ Assess need for additional equipment
___ Hazards to rescuers
    ___ Oxygen deficiency (less than 19.5%)
    ___ Oxygen enrichment (greater than 23.5%)
    ___ Flammable gases or vapors (greater than 10% of LEL)
    ___ Airborne combustible dust (visibility less than five feet)
    ___ Toxic gases or vapors (greater than PEL)
    ___ Mechanical hazards
    ___ Electrical hazards
    ___ Engulfment
    ___ Other: _____
___ Secure area
___ Atmospheric monitoring

Products in confined space:
_____
_____
_____
_____
_____
_____
_____
_____
_____

Notes: _____
_____
_____
_____
_____
_____
_____
_____
_____
_____

	Acceptable conditions	Results			
Oxygen	19.5–23.5%	___ hrs	___ hrs	___ hrs	___ hrs
Flammability	<10% LEL	___	___	___	___
Hydrogen sulfide	<10 ppm	___	___	___	___
Carbon monoxide	<35 ppm	___	___	___	___
Sulfur dioxide	<2 ppm	___	___	___	___
Toxin (specify)	_____	___	___	___	___

___ Ventilation
___ Lockout, tag-out, blank-out procedures
___ Personal protective equipment
___ Air supply (SABA)
___ Communications and lighting equipment
___ Transportation devices for victim
___ Pre-entry briefing on hazards and rescue methods
___ Decontamination
___ CISD

Time: _____ Date: _____ Incident no.: _____

_____
Incident commander signature

Incident commander _____
Safety officer _____
Staging group _____
PIO _____
Police liaison _____
Monitoring group _____
Ventilation group _____
Air-supply group _____
Rescue group _____
Medical group _____
Decon group _____
Entry team _____
Backup team _____
Other _____

This checklist is to be implemented as part of the entry preparation phase.

CONFINED-SPACE MEDICAL CHECKLIST

Date: _____   Incident no.: _____

Entry team   Filled
member's name: _____   out by: _____

_____ Remove personal items
_____ Action plan reviewed
_____ Medical checkout by (see below)
_____ Station uniform removed and protective jumpsuit donned
_____ Air cylinder fully topped off
_____ Communications checked
_____ Proper seal on face piece

ENTRY TEAM MEDICAL CHECKOUT

Pre-entry PSI	Time on air	BP	Pulse	Respiration	Skin

Post entry	Time off air	BP	Pulse	Respiration	Skin

Immediately following:

After five minutes:

Observations noted:

CHAPTER QUESTIONS

1. As a firefighter, what is the most efficient way to prepare yourself and your organization for command of a confined-space operation?

2. What Technician-level member of the rescue team assumes control of the rescue site, coordinating the sectors within it during the course of the operation?

3. What Technician-level member of the rescue team is responsible for the actual placement of teams into the space, as well as the removal of teams and victims?

4. What officer oversees the assembly and outfitting of any personnel who enter the technical-rescue environment?

5. Who should fulfill the role of medical officer during a confined-space operation?

6. Who is responsible for developing any rope-related systems needed for entry or removal of either team members or victims?

7. The essential function of a rescue equipment officer is to act as a _____.

THE INITIAL RESPONSE

Whether you realize it or not, the assessment of any incident begins long before you receive the first alarm. It's hard to imagine a scenario so exotic that certain basics, acquired through planning, training, and experience, won't apply. The initial assessment is critical, for the decisions that you and others make during the first few moments will set the tone for the remainder of the operation. At this point, it is important that you slow down to speed up, or shift into what the members on my team like to call "hyper cool." This is a manner of behaving that all experienced, well-honed fire, rescue, EMS, law enforcement, military, and industrial team members practice. It can't be taught; rather, it is born of confidence and self-discipline, setting you apart from those who exhibit less than exemplary performance during a moment of crisis.

Right about now, your heart is beating so fast that you're sure just about everyone around you can hear it, too. It feels as if it's just about ready to burst out of your chest. Your mouth is dry, and your body is sweating, because in front of you is that situation that you have not only trained for, but even anticipated.

What you need to do is take a deep breath. Internalize whatever stress you may be experiencing. Remain focused like a laser beam, but not to the extent of having tunnel vision. Assess and control the incident. If this incident is like many that I've been to, there will be some confusion, some heightened emotions, and everyone will be looking to you to gain control, maintain control, and bring the operation to a successful conclusion.

By identifying the key components of the situation quickly, you will be able to drive the operation rather than have the operation drive you. This will also provide you with options. The assessment is divided into two distinct concerns, approach and resources, and it is an ongoing, mulilayered process, in that the information gained by each arriving entity is built upon that of those who came before, from first response through specialty units. The approach assessment begins the moment you respond to the alarm. In your mind's eye, and based on your preincident planning, you should already be forming a picture of the incident site. What do you remember of any specific hazards that are present? Are there access problems to the site? What type of space is it? Do you have a response plan or protocol at hand?

Hopefully, your preincident planning has been sufficient enough, and the emergency won't be so exotic, that the situation you ultimately discover will fairly resemble the assessment that you've made along the way. Some emergency scenarios, of course, are downright unpredictable, but if there has been any shortfall in your planning, now is the time that you'll start wishing you'd done a better job. No matter how good the assessment, you should still follow the precepts of the tactical checklist once you arrive on the scene and begin to interview foremen, supervisors, bystanders, ships engineers, and others.

One of the most effective ways of obtaining information about a situation is through the deployment of a reconnaissance team. This team should be assembled as quickly as possible after arrival, meaning immediately. I have seen teams stand around, getting everything textbook perfect, only to discover later that, had they performed a quick recon, they would have found the victim (now dead) right inside the opening of the space.

In this incident, a good recon revealed that the victim was right inside the hatch of the tank truck. Rescuers were able to remove him to safety without the need for an entry operation.

A reconnaissance team should always consist of at least two personnel working according to the buddy system. It goes without saying that they should be wearing the appropriate breathing apparatus and personal protective equipment. Recons can be done by Awareness-level members as long as they know their limitations. Once at the space, they need to address the following questions and relay the information to the command staff.

1. Can you see the victims?

2. Can you make contact?

3. Is the space and its access routes as anticipated?

4. Can the victims be removed quickly with nonentry rescue techniques?

5. Can you determine whether this is a rescue or a recovery?

6. If you're using an atmospheric monitor, what are the readings and detection levels?

7. What is needed at the site right now?

8. What visible hazards exist that require control?

9. What other resources might be needed?

By now, you should have help arriving at the scene. The caliber of the response hinges on the information given by the caller, as well as how the dispatcher interprets that information. An incident that turns out to be a confined-space rescue may, in fact, have been dispatched as an illness call or report of a man down. Thus, a resource assessment is vital toward ensuring that the proper units respond. Of prime concern is how many trained personnel are present and how many more are needed. Summon a technical-rescue team, haz mat team, and specialty equipment as appropriate. If there is any doubt as to whether you will need a particular resource, err on the side of caution and make the request now.

Your actions during the assessment phase are part and parcel of developing an initial strategy. When you complete the assessment, you will have the following information in hand.

1. The number of victims and a potential survival profile.

2. A description of the type, configuration, and access to the space.

3. Any hazards associated with the space or the surrounding areas.

4. Any maps, blueprints, or drawings of the space.

5. A responsible party or facilities engineer who can answer technical questions.

6. Any material safety data sheets (MSDS) regarding chemical products at the site.

At this point, you may opt to perform a rapid rescue and extrication, based on your assessment and the resources at the scene. The Israeli army has a saying that "Sometimes best is the enemy of good," meaning that if you wait for everything to be present and perfect, you may miss a window of opportunity. To give a real-world example, suppose that you have two workers in a manhole. Both are unconscious, and only the first-due assignment is at the scene. None of the specialty equipment has arrived yet. You know that your closest supply of resources is ten minutes away, yet you feel that the victims can't wait that long. Based on your assessment, you can find no indication that a flammable atmosphere exists, but you suspect that an asphyxiant is present. The space is big enough for you to enter and move around without having to take your SCBA off your back, and with a little rope gear that you have on apparatus, another member can rig a quick-removal system using a ladder. You figure that the whole rescue can be accomplished in just about five minutes. Should you go for it? The answer, of course, is yes. Using both knowledge and horse sense, you've made an informed assessment and decided that you can launch quickly with a minimum of risk. We call this "risk-based rescue operations." Among the standard considerations of hazards and victim status is whether or not the benefits accrued from conducting this sort of operation will be conducive to overall success. In the scenario described above, certainly the noxious air was sufficient cause for quick action. Rescue work is an inherently risky occupation. Although we should always be aware of the risks and try to reduce them as much as possible, it would serve no one if bureaucrats and safety advocates were to legislate us right out of business. The only time all of the pieces are in place is in the perfect world of the safety manuals. The standards are good, but not a word in them captures the essence of either what we do or the value of a victim's life.

The hazards of greatest concern are those that OSHA describes as being imminently dangerous to life and health, or IDLH. Such an environment poses one or more of the following threats.

1. An immediate or delayed threat to life.

2. A threat that would cause irreversible adverse health effects.

3. A threat that would interfere with an individual's ability to escape from the space without assistance.

There are products that you will encounter, such as hydrogen fluoride gas or cadmium vapors, that will produce immediate, transient effects that soon pass, even without medical treatment. In many instances, a victim or rescue team member may feel normal, then suddenly collapse and die twelve to seventy-two hours later. Although the effects are delayed, this is still considered an IDLH environment. Although this type of event represents the exotic rather than the typical, it demonstrates the insidious potentiality of hazards. Recognize that, in most instances, it's the common, everyday hazards that will kill or maim you.

There are five major forms of hazards that you will encounter, and you should be cognizant of all of them while making your assessment: atmospheric hazards, burn hazards, mechanical hazards, engulfment hazards, and hazardous materials.

As noted elsewhere, ninety percent of all injuries and deaths in confined-spaces occur as a result of atmospheric hazards. Smell and taste are very poor indicators as to whether a problem exists. In fact, if you rely on them, it may be the last time you do anything. Atmospheric monitors are your first line of defense. Besides the aspects of oxygen deficiency, flammability, and other forms of contamination previously described, it should be noted that hazardous atmospheres can be wayward. The threat may soon exist not only within the confined space, but all around its opening and in the general area as well. Just because the atmospheric concentration of a substance isn't capable of causing IDLH effects, and is therefore not covered under OSHA, does not mean that it doesn't pose a danger to you. There are also contaminants for which OSHA has not yet developed safe-exposure levels. Always assume that the atmospheric conditions within a confined space will change, and you will be one step ahead of the odds.

Monitoring a space using a calibrated, bump-checked, and properly maintained detector is the only way to evaluate the atmospheric hazards present.

Asphyxiation is the leading cause of death in confined spaces. This can occur by way of immersion in vapors or by physical constriction of the victim's chest. Most often, it is due to the quality of air within the space. Asphyxiating atmospheres include those that simply don't have enough oxygen to sustain human life, even if no strictly toxic characteristics are present. Given that the normal concentration of oxygen in air is about 20.9 percent, an oxygen-deficient atmosphere is defined as any parcel of air containing 19.5 percent O_2 or less.

Oxygen deficiency within a space can be caused when oxygen is (1) absorbed by another substance, (2) consumed in a chemical reaction, such as rusting, burning, or curing, or in a biological process, such as bacterial decomposition, and (3) displaced by another gas. This displacement can happen accidentally, such as by the accidental discharge of an extinguishing system or CO_2 system inside a space, or intentionally, such as when a space is made inert with a nitrogen blanket or some other nonreactive atmospheric agent.

POTENTIAL EFFECTS OF OXYGEN-DEFICIENT ATMOSPHERES

20.9%	Normal concentration of oxygen in air.
19.5%	Minimum permissible oxygen limit.
15–19%	Decreased ability to work strenuously; may impair judgment and affect personnel with a history of coronary or pulmonary disease.
12–14%	Increased respiratory rate and pulse rate during exertion; impaired fine motor coordination; perception and judgment altered.
10–12%	Respiratory rate and depth of breathing continue to increase under exertion; judgment becomes poor; cyanosis of lips begins.
8–10%	Mental and judgmental failure, fainting, unconsciousness, ashen face, cyanosis of lips, nausea, and vomiting.
6–8%	Eight minutes of exposure is 100% fatal, six minutes of exposure is 50% fatal, four to five minutes offers a chance of recovery with aggressive treatment.
4–6%	Coma within forty seconds of exposure; convulsions and cardio-respiratory arrest; death.

The atmosphere in a confined space may also be oxygen enriched. Although this is not an asphyxiation hazard, any concentration above 23.5 percent can be a serious fire hazard, since oxygen enrichment can cause combustible materials, including your personal protective equipment, to burn violently.

A flammable atmosphere contains a gas, vapor, or airborne dust at concentrations great enough to burn rapidly on contact with an ignition source, such as heat, sparks, or an open flame. The lower flammable limit (LFL) is the lowest concentration at which the gas, vapor, or dust will be able to achieve sustained combustion. You will often find the terms lower flammable limit and lower explosive limit used interchangeably.

OSHA has established minimum safe levels for the presence of flammable contaminants in confined spaces. To decrease the dangers presented by flammable atmospheres, OSHA requires that the concentrations in confined spaces be maintained at less than ten percent of the LFL, per the alarm settings on monitors.

A toxic atmosphere is one that can cause significant health problems or death. The poisonous physical effects may be immediate, delayed, or compounded over time. Many substances present long-range effects that can permanently disable a victim.

The most common toxic substances found in confined spaces are hydrogen sulfide and carbon monoxide. These, however, are only a few of the hundreds of toxins that you might encounter. Some common sources in the confined-space environment are fuel vapors, tank residue, protective tank coatings, poisons left over from fumigation, the residue of the inerting process, degrading organic material, and even the recirculating exhaust fumes of improperly placed rescue vehicles.

OSHA has developed what are considered to be permissible exposure limits (PELS) for toxic substances. If you find yourself confronted by an environment containing a toxin for which OSHA has not yet developed a PEL, you'll need to do some research. Some sources of information include, but aren't limited to, safety guidelines issued by the National Institute of Occupational Safety and Health (NIOSH), recommended standards published by the National Fire Protection Association (NFPA), manufacturers Material Safety Data Sheets (MSDS), and threshold limit values (TLV) developed by the American Conference of Governmental Industrial Hygienists (ACGIH).

It's important to remember that some products are toxic long before they're explosive, and you'll need to identify those agents early on. Photo-ionization devices are invaluable in this regard.

The danger of toxic hazards in the air may depend entirely on oxygen concentrations and the presence of flammable contaminants. As mentioned above, some atmospheres may be toxic even when they're reduced below flammable levels. Still others when reduced below toxic levels are still flammable. Some substances combined with air pose entirely different atmospheric hazards at varying concentrations. Methane is harmless below a concentration of 10 percent, asphyxiating above 90 percent, and explosive between 10 and 90 percent.

The risks associated with toxic and flammable atmospheres mandate the response of a hazardous materials team with advanced monitoring capabilities during any confined-space operation.

Burns sustained during fires and explosions of flammable atmospheres are not uncommon. Even a brief combustion and flash fire can prove deadly in a confined space. It can sear the lungs and burn rubber hoses on breathing apparatus. The shock wave associated with a blast in a confined space can disrupt hollow organs, resulting in death. Moreover, any such event can consume all of the available oxygen in the space.

Because of the configuration of many spaces, the entrant may not be able to escape or avoid the hazard. Always bear in mind that thermal injuries come not only from fire and explosions, they can also be inflicted by corrosive chemicals, cryogenic components, contact with electric components, and other agents inside a space.

The release of mechanical or electrical energy inside a space can also result in significant injuries or death. Like most hazards, the size and shape of the space may make it impossible for the entrant to avoid coming in direct contact with machinery or electrical components. Still, nearly all mechanical and electrical hazards can be curtailed by employing proper lockout, tag-out procedures. Forgetting to lock out and tag out equipment, or neglecting to take into account backup systems and stored energy in mechanical systems, can result in crush and dismemberment injuries.

Engulfment occurs when someone is immersed in liquid or trapped and enveloped by fine, dry bulk materials, such as grain or sawdust. In some cases, the material doesn't have to be of fine particles. One recent rescue operation in which I participated involved a worker trapped in a hopper containing sizable rocks, two to four inches in diameter, which were going to be crushed for mineral extrication.

The result of engulfment can include asphyxiation from constriction of the chest. A person can also be asphyxiated by aspirating the material. Drowning in a wet environment, such as a well or flooded cofferdam, isn't unusual. Trench operations are closely linked to engulfment scenarios. A victim trapped under several cubic yards of soil, with his head uncovered, may still die from the pressure against his chest, preventing him from breathing normally. Should the engulfing material be corrosive or hot, the resulting burns may be the compounding cause of death.

Working inside a confined space creates noise. Drilling, scaling, grinding, hammering, riveting, and other activities can create deafening conditions within. During a rescue, the movement of equipment, air lines, and air bottles, plus the sheer effort to communicate, all serve to increase the ambient noise levels. Add to this the din of ventilation operations, and the aural problem can, in and of itself, become a significant hazard during rescue operations.

OSHA requires that workers be protected from sound levels greater than 85 dBA for longer than eight hours. Hopefully, a rescuer will never have to be in a space for this long. Still, no matter how long the operation, the noise level must be evaluated and team members must be protected if it is excessive.

Trench operations are closely linked to engulfment scenarios. Here, during a training exercise, shoring jacks and boards are being used to hold back the earth.

Firefighters, EMS personnel, and members of industrial brigades are well acquainted with the effects of heat stress on the body. Even without wearing full turnout gear during a confined-space operation, the vessel or other space may itself become a hazard if the temperature inside rises too high. The sun against a metal tank, limited circulation of the air inside, the insulative qualities of personal protective equipment, and simply the physiological effects of work can all conspire to threaten the safety of a rescue worker.

The safest way to ensure that you are within acceptable limits is by measuring the heat and humidity with a wet-bulb globe temperature indicator (WBGT). NIOSH endorses the WBGT as the preferred method of measuring heat stress and the level of exposure. In the real world, however, I have never seen a rescue team using a WBGT, and the device probably isn't appropriate for rescue operations. The best way to keep heat stress to a minimum is through the physical

120

conditioning of your team members, ample hydration, and proper ventilation. Despite all of this, you will likely be hot, miserable, and spent by the time you exit. Cooling suits and cooling vests, such as those used in Level A haz-mat gear, can be a good option if they are small enough and light enough to be worn in the space.

You can pretty much bet that just about every confined space you work in will have some form of hazardous material or waste. It is imperative that rescue team members have at least Awareness-level knowledge when dealing with haz mats, though Operations-level knowledge as a minimum would certainly be preferred. Incorporating the haz mat team into the response matrix is an important step during the planning phase. Although the subject of hazardous materials is too far-ranging and diverse to be adequately addressed in a text such as this, it is important for the rescuer to bear in mind the types of threats that haz mats pose. The acronym TRACEM (thermal, radiation, asphyxiation, chemical, etiologic, and

Almost every confined space you work in will have some form of hazardous material or waste. Few will be so easily recognized, and you should never ignore the chance that others may be present.

121

mechanical) is useful in this regard. Thermal damage can result from either heat or cold. The skin is the most commonly damaged organ, since it is usually the first to come in contact with the hazard, but other organs can be damaged by prolonged or extreme temperatures. Secondary infection is often a factor. The symptoms of radiation include skin burns, hair loss, and physical abnormalities. Over the long term, they include an increased risk of organ failure and cancer. Asphyxiation is a consideration in any space, but don't be lulled into the belief that the only place you'll find an asphyxiating atmosphere is actually within the space. Chemical burns may present external symptoms, but the internal damage may be more subtle. The function of various organs may be affected by poisoning. The harmful effects may be acquired by a one-time contact (acute exposure), or they may be built up by repeated contact over time (chronic exposure). Etiologic harm is that which traces to a biological cause. Mechanical harm refers to physical damage inflicted by shrapnel or debris associated, for example, with the rupturing of a container or the bursting of a pipe.

All personnel operating as part of a confined-space rescue team should be trained to the haz-mat Operations level, at minimum. You should consider every confined space to be a hazardous materials event until proved otherwise. This means that the haz mat team must answer the alarm as part of the initial response package. Always be alert for placards and warning labels, and learn to read MSDS sheets. Plan appropriately for your response area, and approach every scene with the proper monitoring and sampling equipment.

As with the other disciplines necessary to perform at a confined-space incident, the assessment of hazardous materials and the ability to perform defensive actions is a must. Like rope work, haz-mat knowledge is best gained by means of a haz-mat course, not a confined-space text.

Once you have determined which of these hazards are present, you can prioritize them. The nature and severity of each needs to be considered by the operations officer, and a viable plan must be enacted. This may require pooling the knowledge and expertise of any number of entities. The team deployed to control the hazards must be qualified and well informed.

Determining the hazards before you encounter them is critical to a safe operation. Beyond the problems mentioned above, even everyday occurrences may become a concern. The vibrations generated by pedestrians, city traffic, or passing trains may be enough to compromise a collapse site. Weather conditions or the presence of snakes or spiders inside the space must be evaluated.

One could argue that it would be impossible to identify all of the potential hazards at any given site, and this is likely true. Still, the rescue team should not be faced with surprises that could have been identified beforehand. Evaluating the hazards and controlling them is an ongoing function during any operation. The members must constantly ask whether they have covered all the possibilities and have done all they can to control the threats. Remember, if it's predictable, it's preventable.

Working in a shipboard environment poses its own unique challenges, whether the ship is under way, at anchor, at dock, or in dry dock. Ships present a myriad of hazards and concerns, not the least of which is seasickness. There are also tremendous differences between civilian and military vessels. Whenever you board a ship, you enter a self-contained world that closely combines both living and working environments.

During a rescue operation, there are certain personnel on the ship with whom you must have contact. On a civilian vessel, that would be the ship's master, or captain, if the ship is operational. On the ship, the captain is effectively the law. He must be kept apprised of all ongoing operations. A captain will respect your operations and needs, but you must communicate with him. In some instances, you may be given the first mate or executive officer who'll act as a direct representative of the captain. On a military ship, you must contact one of three persons on board: the officer of the deck (OOD), the command duty officer (CDO), or the ship's fire marshal. The officer of the deck is normally the person in charge of the quarterdeck, or

Ships present a myriad of hazards and other concerns.

that portion of the ship that receives visitors. The command duty officer is the commanding officer's official representative on board. The ship's fire marshal may already be involved in damage control or at the scene of the incident.

Next, you must work closely with the chief engineer. He is an extremely valuable player in any successful operation. A ship's engineer can show you the quickest access routes throughout the ship; advise you of specific hazards; provide blueprints and engineering diagrams; explain the control of mechanical and electrical systems; and generally advise you as to how the ship operates. If you need to know anything about a vessel, the chief engineer or his staff can provide you with what you need.

Because of the very nature of ships, incident command may well be distant from the operations section. The command post may actually be set up on the dock, many stories below and far removed from the actual rescue site. Viable communications equipment, possibly a chain of radio locations, may be necessary to communicate

124

through the steel hull and infrastructure. A hardwired link may be mandatory. Face-to-face communications with the incident commander will be rare because of the distances involved. Most ships are actually high-rises turned on their side and set in water. Consider a shipboard confined-space rescue to be akin to a high-rise fire: manpower intensive and logistically complex, complicated by great distances, difficult access, and long-range communication problems.

Depending on the type of ship that you're working on, you may well be faced by a language barrier. In many instances, foreign ships have multilingual crews, but many other times, no one will speak English. You may need an interpreter, but once again, the initial contact with the captain and key officers is the best route toward identifying and solving the problem.

Obviously, where a ship is located will greatly affect your ability to respond rapidly and assess the situation. If you find yourself having to respond to a ship at anchor or under way, transportation by boat or helicopter will be necessary. In most other cases, you'll be able to drive right to the dock or dry dock.

Given the latter scenario, which is more common, the command post will most likely be set up on the dock, with recon teams and other operational crews entering the ship via gangplanks, walkways, cranes with man baskets, or by some other means. You must decide whether flotation devices are appropriate for the environment you're working in and whether the potential for falling overboard is a real threat to personnel. If it isn't, don't worry about flotation. Because the layout of ships is often complicated, locating the actual space may be difficult and time consuming. If the event occurred below deck, the atmospheric conditions may be compromised. You should always consider this possibility during the initial response and assessment.

Expect your access route on a ship to be a roundabout path—up a gangway, then down one stairway and through a bulkhead, then along a crooked corridor, making a jog to the left, and plunging

Expect your access route onto and through a ship to be a roundabout path.

down another flight. Make sure you have in your company someone who knows the ship well. Sixty-minute SCBA is essential for recon, since you're going to be traveling great distances. Make sure you know where your area of safe refuge is once the decision is made to deploy. Never travel long distances with a mask unless you either have access to SABA or you have pigtail capabilities on your standard SCBA. Each recon team should be equipped with an atmospheric monitor, and that monitor should be on during the entire trek through the ship, not just in those areas perceived to be the danger zones. Additionally, each recon team should be supported by a rapid intervention team. Once you begin your descent into the actual danger zone, be sure to use a marking system to indicate the way in and out. Depending on the vessel, many options are available. Fire line tape, anchored at the point of entry and stretched along the ingress route, is useful. Search rope is another option, but fire line tape is cheaper, it's easier to manipulate, it's visible in the dark, and it won't hang you if you happen to run into it.

A recon is a sneak-and-peek mission. Its purpose is not one of rescue, unless you happen to come upon someone who can be quickly removed. Get in, assess the site, and get out. Relay the information to the IC and commence the larger operation.

CHAPTER QUESTIONS

1. How quickly should a recon team be assembled and deployed?

2. What is the importance of a resource assessment made early during the course of an incident?

3. Sometimes best is the enemy of _____.

4. An IDLH environment poses one or more of what three basic threats?

5. What are the five major forms of hazards that you may encounter in a confined space?

6. An oxygen-deficient atmosphere is defined as any parcel of air containing _____ or less.

7. Consider any oxygen concentration above _____ to be a serious fire hazard.

8. The two most common toxic substances found in confined spaces are _____ and _____.

9. What is a PEL?

10. OSHA requires that workers be protected from sound levels greater than 85 dBA for longer than _____.

11. What are the threats posed by hazardous materials, as referenced in the acronym TRACEM?

12. What entity on a ship may best show you the quickest access routes; advise you of specific hazards; provide blueprints and diagrams; explain mechanical and electrical systems; and genrally advise you as to how the ship operates?

CHAPTER EIGHT

Operations Prior to Entry

A fter gathering all the information you can and completing the initial assessment, you will be able to determine whether the operation should proceed in a rescue or a recovery mode. If the first responders at the Awareness level have done their jobs, the hot, warm, and cold zones should already be established. As in any incident, the hot zone is the area of greatest hazard, including the confined space itself; the warm zone is a restricted area surrounding the hot zone, into which only personnel participating in the operation may be permitted; and the cold zone is where the command post, staging, rehab, the media, and bystanders are located.

As mentioned earlier, all energy sources and machinery must be isolated and locked out, and the space must be deemed safe from such hazards before the team can commit to the entry. Be sure to follow all applicable procedures in doing this. Never cut hydraulic lines. Electrical lockout is typically accomplished by throwing a main switch in a circuit box, or even by flicking off a wall switch. Lock off the main as appropriate or, as in the case of a wall switch, place a lockable box over it, complete with a warning tag.

Depending on the SOPs of your department, each entry member may keep his key to the box; otherwise, the operations officer may collect all of the keys and reissue them when personnel are exiting the space.

In cases where material is being conveyed through pipe, it may be necessary to blank the pipe. This is done by fastening a plate, known as a skillet, to cover the bore. The skillet must be able to withstand the highest pressure that will be exerted against it. The disadvantage of this method is that the pipe must be unbolted at a flange, requiring time and knowledge. Someone must be positioned upstream of the break to control the flow at a valve, if that's even possible. It may also mean that you'll have to deal with whatever residual fluid comes out of the line, which can be a dubious proposition, depending on the contents of the pipe.

Another method, double-block and bleed, allows you to render a length of pipe safe by locking out and tagging a vent valve located between two other in-line valves that have been closed off. In most instances, double-block-and-bleed systems are installed in piping that requires frequent shutdown for cleaning or other maintenance. It may be possible, in some instances, simply to sever the pipe after closing an in-line valve. In a rescue operation, this may well prove the most practical and expeditious method. It may be an unacceptable option, however, depending on the product within.

Isolation of the space is usually accomplished by a hazard-control team. Such a team might merely consist of a single member escorted by an industrial representative; it could also be an entire engine company.

Entry depends on constant atmospheric monitoring. All monitors are designed to be used for one specific gas, and each is calibrated to that particular gas. This makes a monitor only accurate for that gas, but that doesn't mean you can't detect others. All monitors have a relative response curve. As calculated by the manufacturer, this curve indicates the response of the monitor to other gases. The curve provides a series of correction factors, typically expressed in

ATMOSPHERE	LEVEL	ACTION	MONITOR
Combustible/ flammable gas	10% of LEL	If outside the space, mitigate the problem. If inside the space, exit.	Alarms both visually and aurally.
Oxygen	Less than 19.5% or greater than 23.5%	If outside the space, determine the cause of the problem and correct it. If inside the space, exit.	Alarms both visually and aurally.
Toxicity	CO–35 ppm H_2S–10ppm	If outside the space, determine the cause of the problem	Alarms both visually and aurally.
PID levels	Background to 5 ppm.	Use Level C PPE.	Alarms both visually and aurally.
	5–500 ppm	Use Level B PPE	
	500–1,000 ppm	Use Level A PPE	

tabular form. If your monitor was calibrated for pentane, for example, and the display is indicating 70 percent of the LEL while monitoring ethane, with a correction factor of 0.7 on this instrument, then your actual reading would be 70 X 0.7 = 49, or 49 percent of the LEL. Some monitors have multiple relative response factors built into them. One automatically makes calculations based on the relative response factors for over twenty-nine gases.

Applying the information you gain from your atmospheric monitor is a matter of writing action guides, or SOPs, for different contingencies. The above action guidelines should be used for confined-space operations.

In order to make sense of the readings and apply what they mean, you must know how your monitor works and what its limitations are. In Chapter Five, we discussed how various monitors work. Standing at the entrance to the space, monitor in hand, how do you use that instrument to obtain the best readings possible?

Because of vapor density, you should always monitor at three different levels within a space. If, for some reason, you only measure the bottom third of a space, you may be missing at least two or three contaminants above. Methane is lighter than air, and it will rise to the top of a space. Carbon monoxide weighs about the same as air, and it will tend to collect at a middle height or else disperse throughout the space. Hydrogen sulfide is heavier than air and will find its way to the floor. To read these or any other gases accurately, you must be aware of the limitations of your device. Understanding the relationship between flammability (expressed as a percentage of total volume) and toxicity (expressed in terms of parts per million) is essential. Each 10,000 ppm equals one percent of concentration. Unfortunately, relying on the toxic sensors inside your monitor or attempting to use the LEL sensor to convert to toxicity can be dangerous. Your monitor can only detect down to 10,000 ppm, or one percent. Many products are toxic at levels well below that, and photo-ionization devices are becoming much more popular for capturing the necessary information.

Vapor pressures are also important. Substances that have a high vapor pressure, meaning that they produce vapors, are more dangerous, since they'll move throughout a space, seeking any outlet they can find. Any number of substances with high vapor pressures are either toxic, flammable, or both. Some liquids also move readily from the liquid to the gaseous state, and this is a direct function of vapor pressure. The higher the vapor pressure of a liquid, the more likely it is to evaporate. Like atmospheric pressure, this characteristic is usually measured in millimeters of mercury. You can research materials to find their vapor pressures. For purposes of comparison, the vapor pressure of water is 25 mm Hg. Acetone ranks at 250 mm Hg, and acetylene at 2,500 mm Hg.

Predictably, it is important to know the operating parameters of your monitor, such as how long the sensors will last and whether the monitor is RF shielded so that it won't be affected by electricity and radio waves. If you're using a monitor with a hand aspirator, find out how many pumps are required to bring a sample through the length of the tube. A water filter on the end of the intake tube may be necessary. The technical manual will help you answer these and other questions. Follow the manufacturers recommendations for calibrating the device in the field.

A basic rule of using all monitors is to take your first sample from a small opening before opening up the space, as well as to stand upwind of the location. Many times, the mouth of the space is already open. Still, remember that high concentrations of toxins and flammable gases can accumulate under covers and around openings. Although gases can be insidious, avoiding pockets and releases of highly concentrated gas is often a matter of common sense.

Assuming that you have a four-function monitor capable of measuring oxygen, combustible levels, toxicity, and pH, there is a hierarchy to follow when dealing with confined spaces. The order of monitoring should be (1) pH, (2) oxygen, (3) flammability, and (4) toxicity.

Measuring pH is first on the list for good reason. Before you even consider inserting that nice, new, $2,500 probe into the hole, you'd better do a quick check of the corrosive potential within the space.

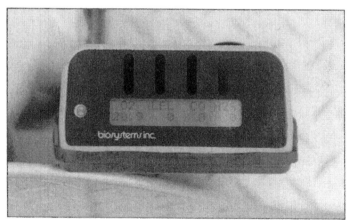

When using a four-function monitor, measure pH, oxygen, flammability, and toxicity, in that order.

Measuring pH can be done by using litmus paper or meters. Since litmus paper costs about $2.50 a roll, it really makes sense to save the monitor. The pH scale runs from 1 to 14, with 7 at the center. Values lower than 7 indicate an increase in acidity, while those higher than 7 indicate increasing alkalinity. The scale is logarithmic, meaning that the difference between a substance with a pH of 6 and one of 7 is small, whereas the difference between 3 and 4 is much greater.

The monitor with which you're working probably measures oxygen concentrations between 0 and 25 percent. Since you're using it for confined spaces, you should set it up to alarm at 19.5 percent, which is the minimum permissible percentage established by OSHA, and it should also be set to alarm at levels above 23.5 percent. Be aware that your ability to use any monitor depends on certain limitations. Monitors are affected by altitude, for one thing. At high altitudes, you'll get a low reading for oxygen, and at low altitudes, you'll get a higher reading. The sensor has a shelf life, and it may not be any good if you haven't checked it lately. If you're in a really cold environment, the electrolyte might freeze. If you didn't verify the pH of a corrosive environment first, you may have ruined the electronics and the unit won't work.

In checking for flammability, you are primarily determining the presence of hydrocarbon products that might be present in the space. Certain instruments are designed to measure methane only. These measure the flammable vapors as a percentage of the lower explosive limit. The monitor that you're using is calibrated to a certain flammable gas, either methane, pentane, butane, or hexane. There are four basic scenarios under which you should always conduct a flammability test: in any space that you suspect is contaminated, before any entry into a confined space, to detect potential leaks, and if you are investigating an unknown material.

The monitor that you're using has a preset alarm level for 10 percent of the lower explosive limit. This means that, when the level of the product you're testing reaches 10 percent of its LEL, the monitor will sound an alarm. This level, one-tenth of the LEL, is an action level, the level at which you need to make a decision. It isn't a panic level. The explosive limits for methane, for example, are at 5 percent lower and 15 percent upper. Beyond these limits, the mixture is either too lean or too rich to burn. If your alarm goes off at 10 percent of the LEL, you are actually in an atmosphere of only 0.5 percent methane, or 5,000 ppm.

Remember that we tested for oxygen before taking a read on flammability. This is because monitors must have a certain percentage of oxygen present so as to function properly. Most instruments require an atmosphere of between 14 and 16 percent O_2 to give an accurate reading. Assessing the oxygen levels first will give you an indication as to whether the flammability readings are going to be accurate.

Additionally, most monitors can't be used in atmospheres containing silicon products, sulfur compounds, acidic compounds, and leaded gasoline. Also, certain aircraft fuels, such as JP10, won't register on some monitors. You also won't be able to detect flammable mists or dusts, and always be aware that the response time of the instrument is highly dependent on the ambient temperature.

Since methane is such a common gas, we often expect it to be present in areas such as vaults, sewers, and any underground location where organic material may be present. If you're measuring an unidentified gas, it might be appropriate to use a charcoal filter to determine whether methane is the culprit. The methane molecule is so small that it will pass through charcoal, whereas other gases won't.

LEL sensors were originally designed to solve the problem of measuring methane in mines. Most LEL sensors use a Wheatstone bridge to measure the change in resistance resulting from a flammable gas burning on a heated wire. Essentially, an LEL sensor measures the temperature at which a particular gas burns. That temperature is converted to a percentage of the LEL. However, some gases burn hot, whereas others burn relatively cool. These differing physical characteristics lead to some significant concerns when using LEL sensors to measure all but short-chain saturated hydrocarbons, such as methane, propane, and ethane. For example, gasoline burns at approximately half the temperature of methane. Thus, it produces half the output on the display of an LEL meter. If an LEL monitor is calibrated to methane and then used to measure gasoline vapors, the display will give only half the true value. An atmosphere that registers at 50 percent of the LEL would actually be at the 100-percent point.

The toxicity of a substance is usually determined through animal testing. You may see the measurement LD_{50}, which is frequently used. This is the amount of a substance that, on exposure, will kill half of the animals in the test.

The monitor that you're using has either one or two toxic sensors, most likely set up to measure hydrogen sulfide (H_2S) and/or carbon monoxide (CO), the two most common toxins. The alarm settings are at 35 ppm for carbon monoxide and 10 ppm for hydrogen sulfide. As before, these are action levels, since they represent time-weighted averages set by OSHA for an eight-hour exposure. The sensors for these two gases contain an acid electrolyte solution that may be neutralized by alkaline vapors or other gases. Interestingly enough, one of the gases that may neutralize a CO sensor is hydrogen sulfide. Most CO and H_2S sensors are now manufactured or filtered to eliminate the interference caused by

EFFECTS OF HYDROGEN SULFIDE EXPOSURE

ppm	Effects and Symptoms	Time
10	Permissible exposure limit (PEL).	8 hours
50–100	Mild eye and respiratory irritation.	1 hour
200–300	Marked eye and respiratory irritation.	1 hour
500–700	Unconsciousness, death.	$1/2$ hour
1,000	Unconsciousness, death.	Within minutes

EFFECTS OF CARBON MONOXIDE EXPOSURE

ppm	Effects and Symptoms	Time
35	Permissible exposure limit (PEL).	8 hours
200	Slight headache, discomfort.	3 hours
400	Headache, discomfort.	2 hours
1,000–2,000	Confusion, headache, nausea.	2 hours
1,000–2,000	Heart palpitations.	$1/_2$ hour
2,000–2,500	Unconsciousness.	$1/_2$ hour
4,000	Death.	Less than 1 hour

other gases, but some are not. Some of the gases that may interfere with these sensors are acetylene, dimethyl sulfide, ethyl alcohol, ethylene, sulfur dioxide, propane, mercaptan, nitrogen dioxide, methyl alcohol, isopropyl alcohol, and hydrogen cyanide.

The physiological dangers of exposure to these two gases are expressed in the tables above.

As mentioned above, the most accurate way to measure toxicity is with a photo-ionization device (PID). This device will provide you with measurements of the largest group of compounds, known as the organics, or compounds containing carbon molecules. These substances include aromatics (benzene, ethyl benzene, toluene, xylene); ketones and aldehydes (acetone, methyl ethyl ketone, acetaldehyde); amines and amides (carbon compounds that contain nitrogen), chlorinated hydrocarbons; sulfur compounds; saturated hydrocarbons (butane, octane), unsaturated hydrocarbons, and alcohols. Additionally, a PID can measure some inorganics, such as ammonia. Still, a PID can't measure radiation, nitrogen, oxygen, carbon dioxide, water vapor, carbon monoxide, hydrogen cyanide, sulfur dioxide, natural gas, methane, propane, ethane, acidic gases, or freon. Therefore, this device should only be viewed as one component in a system.

Monitoring should take place prior to entry and preferably continuously until the event has been declared terminated. Some SOPs indicate intervals of five to fifteen minutes at the entry point anytime personnel are inside. This is not a hard-and-fast time frame, and it should be adjusted based on the severity of the atmosphere with which you're dealing. Concerns about dangerous or changing atmospheres should prompt you to monitor more frequently. One person should conduct the monitoring and record all of the data, and this should be his sole task throughout the operation. That person should never leave the monitor unattended. All readings should be reported to the extrication officer or the operations officer, and any fluctuations or severe changes should be reported immediately. Additionally, the readings acquired at the mouth of the space should be compared, on an ongoing basis, with those acquired by the entry team.

Atmospheric monitoring is a crucial component of any tactical decisions made during the course of an operation. It is a tool that identifies hazards and provides you with a baseline from which to proceed. Still, to know that a specific hazard exists is only part of the puzzle, and although it gives you an opportunity to protect yourself, it does nothing to eliminate that hazard. In this regard, a number of options may be available to you. Some may be easy and some more difficult to implement. Whichever you choose, you should continue to use atmospheric monitoring as a guide for decision making and as a benchmark to determine whether your strategy to control the hazard is working.

If the problem is an atmospheric one, it makes sense that you might be able to control it with ventilation. This method of mitigating a hazard is only as good as the technique you use to employ it, and it is further limited by the configuration of the space, the type of ventilation you use, and the nature of the contaminant. If you initiate action when the substance has reached the alarm level, yet it shows no sign of dropping to acceptable levels, you may not immediately be able to determine the cause. Is it because the ventilation isn't working? Is it because there is some liquid product that continues to emanate vapors despite all your efforts? At some point, you'll need to make a decision. Should you continue ventilating, possibly increasing the concentrations to their flammable range? Or should you perhaps allow the substance to rise above its upper explosive limit to where it can't burn, then decide on another strategy? Ventilation is the method of choice during rescue operations, since it is fast and easy to monitor, but you can't expect it to work every time.

It has been stated before that hazardous atmospheres cause the most fatalities in confined spaces. Firefighters are well acquainted with the benefits of using ventilation at structure fires. Although

Although the principles are the same, it is often much more difficult to ventilate a confined space than a structure fire.

139

the principles are the same, it is often much more difficult to ventilate a confined space than a house on fire.

The purpose of ventilation in a confined space is to replace the oxygen-deficient, flammable, or toxic environment with clean air, and ultimately to eliminate the atmospheric hazards. Tactically, ventilation is one of the first tasks that should be accomplished, and it can often easily be accomplished by first responders at the Awareness level of training. The vast majority of ventilation procedures enacted prior to entry involve fans. There are a variety of tactical applications for these devices, each dependent on the nature of the event, the nature of the contaminant, and the configuration of the space. To be successful, you may wind up employing two or more methods during the course of any given event.

There are two forms of mechanical ventilation, positive and negative, meant to supply air or exhaust air from the space. The fans that you purchase for confined-space operations are capable of providing both. Positive-pressure ventilation, also called supply ventilation, works by mixing clean air with the contaminants, thereby diluting them. If there are adequate openings situated in the right places, it may also drive the contaminants right out of the space. Positive pressure may create a new hazard by pushing contaminants to other parts of the space or by agitating the regions within and driving contaminants into nearby areas.

In contrast, negative-pressure ventilation draws contaminated air out of a space. Fresh air then enters through any available openings to replace the air being exhausted. The advantage that this method has over positive pressure is that it allows you to control the direction of the inward flow by means of ductwork. A good example of this would be a bulk storage tank containing heavy petroleum vapors that have accumulated near the bottom. Negative-pressure ventilation would work best if the ductwork were designed to conduct the incoming fresh air to the lower levels, where the vapors are the most intense. If the product that you intend to draw through a fan is above its LEL, you need to ensure that you have eliminated all sources of ignition, including sparks from static electricity. You might consider bonding and grounding your equipment in extreme circumstances.

These two means of mechanical ventilation may be used in combination. Positive/negative-pressure ventilation, otherwise known as push/pull ventilation, can be a highly efficient way to move air and its contaminants from within a space. You must ensure that you aren't dumping harmful vapors into a location that will affect the operation or jeopardize others.

Mechanical ventilation can help eliminate atmospheric hazards, but you can't simply turn on the blower and go to work. It takes time to replace a contaminated atmosphere. For rescue operations, we are simply concerned with creating an environment that is within acceptable working limits. You should never assume that ventilation is an acceptable substitute for atmospheric monitoring. The contrary is true, that you should base your ventilation needs and approaches on the results of atmospheric monitoring.

The essential goal is to reduce flammable atmospheres below 10 percent of the LEL, reduce toxic contaminants below the PEL, and increase oxygen levels to above 19.5 percent. You won't always be able to do this, but neither will precarious conditions always preclude your entering a space. The decision to enter is predicated on tactical decisions, as well as the overall mission. If you're operating in the recovery mode, you can afford to make the atmosphere pristine. If you're involved in saving a life, you will take the best conditions you can get, weigh the risks, and live with your decision.

All of the information that you can acquire regarding the physical properties and exposure risks of a substance will influence your choice as to what specific ventilation technique to use. The source is also important. A point source, such as a leaking valve or broken pipe, will generate a radiating hazard whose highest concentrations are in the regions close to the source. In such a situation, you might apply negative-pressure ventilation, attempting to draw from the point of the leak and venting this air to the outside. A cofferdam or solvent tank in which the product is covering the bottom tends to present a scenario in which the hazard is spread more uniformly over a wide area. In such a case, positive/negative ventilation is likely to work best. Always consider whether increasing the airflow within a space might create a fire or explosion hazard that didn't exist before you disturbed the status quo.

The spaces that you encounter will vary in volume and shape, as well as the number of openings and internal obstructions. Since smaller spaces are so much easier to ventilate, often you can use a quick estimate of volume to help you select the most appropriate method.

Even when done properly, it will take time to replace a hazardous atmosphere with a tolerable one.

Purists will note that the efficiency of a fan may be calculated by a simple mathematical formula, $Q = a \times V$, in which Q equals the volume of airflow in cubic feet per minute (cfm), a equals the cross-sectional area of the duct or blower in square feet, and V equals the velocity of the air through that duct in linear feet per minute (lfm). In the real world, the manufacturer has already rated the efficiency of the fan for you, and, short of finding the lfm by either direct measurement or by backtracking through the equation, you won't have a reliable number for this variable. (Simply backtracking through the equation will likely arrive at an erroneous value, anyway, given the physics of airflows, especially in the presence of restrictions. The exact output of the fan also depends on a host of variables.) Ultimately, the lfm value is not at issue, anyway (unless an airstream is so restricted that it can't meet the demands of the fan—a situation that you should have already recognized before resorting to math). In the real world, a rescue team will initiate ventilation with whatever fans it has available, monitor the results, and call for more fans as appropriate.

Of more value in the field is to calculate the number of air changes per minute that your ventilation fans can provide. To calculate this, divide the total cfm of the fans in operation by the volume of the space, expressed in cubic feet. Suppose you have a fan rated to move 3,000 cfm and that you are using it to clear the air, without restrictions, in a 30' X 30' X 30' space.

Volume of the space = 30 X 30 X 30 = 27,000 cubic feet.

Blower capacity = 3,000 cfm.

Air changes $= \dfrac{3,000}{27,000} =$ 0.111 air changes per minute, or one air change every nine minutes.

It usually takes about ten to fifteen air changes to flush the air-borne contaminants from a space, provided that no other source still exists within. Allowing for fifteen air changes in the above example, it would take this ventilation equipment two hours, fifteen minutes to flush all of the contaminants from the space.

The throw of a fan, usually calculated by the manufacturer, is also important in gauging just how far away from the fan or duct the ventilation will be effective. A rule of thumb is that 200 lfm is required to mix air and move contaminants. Generally, when using positive pressure, the velocity of air at a distance thirty diameters away from the fan or duct is about 10 percent of the velocity at the face of that fan or duct. If you're operating in a negative-ventilation mode, throw is reduced far more, falling to 10 percent of the face velocity only one diameter away from the fan.

Predictably, the addition of ducts reduces flow. Make sure that you are aware of your fan's limitations by consulting the manufacturer's literature.

In designing the ventilation system to suit the incident, the first problem to watch for is the recirculation of exhausted, contaminated air back into the space. The second major problem is that of short-circuiting the airflow. This occurs when fresh air moves directly from the air inlet to the exhaust outlet without circulating through contaminated areas of the space, meaning that much of the space will never get ventilated. Guarding against these faults involves the proper use of equipment in the right locations. Ductwork may be required. Ideally, you should have incoming fresh air and the exhaust air moving through separate openings, but the typical confined space won't leave you with much choice. Most confined spaces have just one opening, and even if you're lucky enough to have more, they probably won't be in the most advantageous locations. This problem can be solved, at least partially, with ductwork. You should place the ducts in such a manner that they won't be damaged when you're working around them, and they should be as straight as possible. Make sure that all connections along the way are tight. At the end of a duct, use a wall or some other flat surface as a diffuser, deflecting the airflow into corners and to all levels within the space.

In some instances, when you are unable to change the atmosphere as quickly as you would like, you may opt to create an air tunnel, opening up a lane between the rescue team and the victim. This won't eliminate the hazard, but the pressure of the incoming fresh air will push the contaminants to other sections within the space. Although risky and not a primary option, this may prove to be your only chance at pulling off a rescue. It may be possible to replace one atmosphere with another, a technique that should only be employed during a recovery operation. Purging the space with nitrogen or carbon dioxide for a recovery operation will displace the toxic or explosive atmosphere with an asphyxiant, but at least an inert atmosphere is preferable to a volatile one. In contrast, sometimes the only option is to pump out the product, such as when confronted with tanks containing sewage, chemicals, or fuel.

Previously we discussed in general terms the personal protective equipment that a rescue team should have available. Now you must make a decision as to what gear the team will wear on entry. You want to protect your personnel from the worst-case scenario, based on your assessment. Although no one type of PPE can protect against all the hazards, there are some guidelines for you to follow.

First, we should discuss what isn't appropriate for use in confined-space operations. The limitations of space pretty much rule out structural firefighting gear except for those members operating outside the space. Does this mean that you absolutely cannot enter a space wearing turnout gear? No. It only means that your profiles for heat stress, immobility, and inability to function will increase dramatically if you choose this option. Turnouts are appropriate for the outside, but not the inside. Additionally, short-sleeve shirts, tennis shoes, blue jeans, cotton jumpsuits, and fire service helmets are all unsatisfactory for entry operations.

Basic protective attire is as listed below. Recognize that this minimum of apparel will not allow a team member to come in contact with any chemicals or biological hazards.

- Long-sleeve, flame-retardant outer wear.

- Low-profile helmet with chin strap.

- Inner fire-resistant flight glove.

- Outer leather work glove.

- Fire-resistant hood.

- Steel-toed leather boots.

- Knee and elbow pads.

- SABA.

- Personal alert device.

- Hand light.

Structural firefighting gear is generally appropriate only for those members operating outside the space.

146

- Atmospheric monitor (one per entry team).

- Class III harness.

- Communications system.

- A fall-arresting and belay line for vertical spaces of greater than four feet.

- Air supply for the victim.

A number of manufacturers make protective jumpsuits out of the same material used for turnout gear, giving the wearer Level II protection. This type of garment was originally designed for the U.S. Navy for shipboard firefighting, and it will offer an additional amount of flash protection when needed. It will also add bulk and weight to the rescuer, and it will increase the likelihood of heat stress.

In cases where team members will be exposed to chemicals or biological hazards, additional protection must be worn over the basic ensemble. Two types of garments are available as a defense against chemicals. A vapor protective suit, also known as a Level A suit, will provide the highest level of protection. Resistant to chemicals, it is worn when the substances encountered are volatile or have significant toxicity and absorption characteristics. Typically, such a suit totally encapsulates the wear. Even the breathing apparatus is worn inside, providing a gas-tight environment. Level A suits come in both reusable and disposable varieties.

Operating in an environment that requires this level of protection will pose a few related considerations. If the chemical is that bad, then the object of your mission, the victim, is almost assuredly dead. The operation will therefore be a recovery, and there may very well be a better way to recover the body than to enter the space in a Level A suit. In twenty-three years in the fire and rescue services, I have never seen a Level A entry into a confined space. Haz mat environments, yes; confined spaces, no. These suits are difficult to move around and work in, and if you tear one in a hostile environment, you'll be in deep trouble. Still, chemical protective gear shouldn't be used alone when you are faced with a heated or flammable atmosphere. Should you find yourself in a flash, you'll be shrink-wrapped in the suit. You must meet certain minimums in the HAZWOPER guidelines (Hazardous Waste Operations and Emergency Response; OSHA 29 CFR 1910.120) to wear one of these suits, meaning that you must be a haz mat technician or you risk violating federal law. Many departments require more than the standard forty-hour HAZWOPER course for haz mat technicians. Expecting Technician-level personnel to maintain all of their certifications in tech rescue and haz mat is asking a little much, however, and the result will invariably be a lessening of competence within the team for any given skill.

A splash-protection suit, also known as a Level B suit, is designed to keep liquids off the wearer's skin when used with the proper boots and gloves. Although reasonably liquid tight, they should by no means be used in cases where immersion is possible. At their openings, they are not impervious to gases. Still, Level B suits are much more appropriate in the confined-space environment when dealing with certain liquid chemicals, or bio hazards such as sewage. As with vapor-protective suits, these garments will melt or burn if exposed to flames.

As the members of the rescue team assemble and equip themselves for the mission, it is imperative that you brief them prior to ingress. This briefing can take place at the point of entry or while the members are dressing and running last-minute checks. It is usually conducted by the operations officer in conjunction with the extrication and safety officer. You should always include a diagram, map, or drawing of the space that the team is about to enter. Each member should come away knowing exactly what his tasks will be and what is expected of him. Any safety considerations must be a component of this briefing. This will entail not only a prep on the hazards to be expected, but also reminders about relevant operating practices. Each team should be advised of the time limits enacted by the extrication officer. If, while in the hole, any members find that they are unable to accomplish a specific task or need additional time, they should request an extension and state how long they estimate that it will take to complete the task. The safety talk is the juncture at which the members should run buddy checks on one another, ensuring the integrity of their personal protective equipment. Besides an appraisal of the atmospheric readings and a description of the entry method to be used, emergency procedures should be put in force. At the end of the briefing, run a final check of all communications equipment.

By now, you should know what chemical or biological hazards you're facing. Based on the threat, you may need to get your haz mat team to establish a decon area in the warm zone. In most cases, the process will be one of gross decon, or a three-step process to remove any residual material that the entry team brings out. You may also have to decontaminate the victim before the forward medical team can receive him. This is highly dependant on the victim's condition, of course. There simply may not be enough time

By the time a technician enters the hole, he should know exactly what his tasks will be and what is expected of him.

to decontaminate him thoroughly before the initiation of definitive treatment becomes mandatory. Always remember, it's better to be dirty and alive than clean and dead. These determinations are best left to a trained haz mat team with the research material and resources to provide proper care to team members and victims. Either way, be sure to notify the medical control hospital well in advance as to whether you'll be delivering a decontaminated or a partially clean victim.

Decontamination is a complex issue, and a number of questions always apply. Which personnel will perform the decon, what level of protection they should wear, where the decon line should be located, and what sort of cleaning solution to use are all substantial questions. Beyond that are matters of even greater delicacy, many of them environmental. Where will the water supply for decon come from? How will the contaminated runoff be handled? Is a cleanup contractor required to remove the waste after testing? Will equipment need to be bagged and disposed, or can it be taken to another location for additional cleaning? What type of medical baseline screening needs to be done after decon? If you are unsure about the answers to these types of questions, a skilled haz mat team should be able to provide you with most of the information you need. In a pinch, the removal of clothing and an emergency decon with copious amounts of water will begin the process in lieu of any other action.

Part of the entry team should be made up of cross-trained technical rescue physicians and paramedics. With frontline medicine available in the space, the remainder of your contingent should establish a forward medical treatment and screening area. Although there may not be enough cross-trained personnel to staff this function fully, at least one technical rescue paramedic or physician should supervise the medical staff assigned to this location.

Similarly, you should set up a forward logistics site. Although the main location of resources is usually at the base, an area remote from the actual rescue site, confined-space operations generally require a forward logistics site to support the entry. In some instances, a single rescue equipment officer can run this site; in other instances, a staff of three or four might be required. A forward logistics site for a confined-space operation should have enough equipment and supplies on hand to address the needs of the operation as it progresses. A minimum inventory is listed below.

- Air bottles for supply carts or breather boxes.
- Extra breather boxes or supply carts for the SABA.
- Extra hose for the SABA.
- Extra SCBA.
- Full-body protective suits.
- Extra radios, if possible, and extra batteries, if applicable.
- Rope, webbing, harnesses, and hardware.
- Power cords, junction boxes, and extra generators.
- Extra fans and accessories.
- Tool box.
- Air supply repair kit, including extra fittings, adapters, tape, and the like.
- Portable compressor, if possible; otherwise, a high-pressure line run from a remote site.
- Backup retrieval system.

Confined-space operations require tremendous quantities of air. There are a number of tasks that the air-supply officer and the personnel assigned to him need to accomplish before the rescue team can enter the space. Both the primary and secondary air-supply units must be at the site and ready to become operational. This includes one supply system for each team and backup team, as well as an emergency backup system in case the primary should fail. Some method should already have been established to refill the air bottles for the supply system. This can be accomplished by shuttling bottles to a remote site for refill, or perhaps by means of a portable compressor, cascade system, or a high-pressure line from a remote cascade.

There must be adequate supply lines for the SABA at the site and ready to go. In prepping the air supply lines, there should only be couplings at the user and supply end connections. All other connections should either be screwed together or otherwise fastened so that they will not part company. This will lessen the chances of snagging a coupling during entry and having the line come apart.

Actually, the management of air hoses begins long before the team enters the space. All air lines should be color coded, either by using hoses of different colors or by marking them with tape, by painting the connections, or by sheathing them in webbing. This allows the inside and outside teams to identify and control the lines even when they can't see each other, and when multiple lines are run into a space, that's a critical concern. When deploying them, each section of line should be laid out and readied, and you must ensure that you have enough length for entry. Naturally, you should keep a kit full of air-supply hardware on hand, stocked with such items as extra hose, bottles, air fittings, duct tape, and the like.

CHAPTER QUESTIONS

1. In locking out equipment, when is it appropriate to cut a hydraulic line?

2. What is double-block and bleed?

3. What is a relative response curve?

4. What characteristic of a gas mandates that you always monitor at three different levels within a space?

5. What is vapor pressure?

6. When using a four-function monitor, the testing of a confined space should be conducted in what order?

7. True or false: Hydrogen sulfide may neutralize a CO sensor.

8. Photo-ionization devices measure what group of compounds?

9. Among other objectives, the essential goal of ventilation is to reduce flammable atmospheres below what percentage of the LEL?

10. How many air changes are usually required to flush airborne contaminants from a space, provided that no other source exists within?

11. True or false: A Level B suit is appropriate protection in cases where immersion in a liquid contaminant is possible.

12. Who usually conducts the preentry briefing?

CHAPTER NINE

ENTRY AND RESCUE OPERATIONS

Getting the teams in and out is the duty of the extrication officer. It is he who is immediately responsible for assembling, outfitting, and deploying the rescue and back-up teams during any confined-space operation. By careful delegation of assignments during the pre-entry phase, this officer puts into place everything needed to get the team in and out of the space, and he ensures the safety of the site they are working in and around.

In establishing the rescue team, the extrication officer is instrumental in determining the viability of the operation. Many times in this discipline, choosing the right person to go into the hole is simply a matter of conscripting the smallest member of the team. Other times, it means selecting personnel for their skills and expertise in a given area. In putting together the entry team for a rescue, the extrication officer must always consider the welfare of the victim. If you expect delays in removing the victim, at least one of the entrants should be from among your technical rescue paramedics or physicians.

Team members should always work in pairs. Given OSHA's two-in, two-out rule for imminently dangerous environments, this means that you must always have two additional personnel outside ready to assist in the event of an emergency. This does not mean, however, that you must have two additional members outside for every two inside. A total of two outside is all that is required, but of course, your rapid intervention team should be formed according to the task at hand, and a larger contingent may be necessary. A RIT's sole purpose is to rescue rescuers. If you deploy the RIT to assist with the extrication, get another RIT in place at once.

Sometimes choosing the right person to go into the hole is a matter of conscripting the smallest member.

Each entry team needs its own radio designation, whether Entry 1, Entry 2, or some other means of address. In some instances, you might only have one hardwired communications system, in use by the primary team. If so, you'll need to decide on an additional system for the backup and rapid intervention team. This could entail either radios or another hardwired set. Each entry team also needs its own air supply, meaning a separate air cart, breather box, or some other supply system. If you depend on one system and that goes down, neither the RIT nor the backup team will be able to make the entry. Additionally, four members breathing off of one source of supply will put a strain on both the air-supply system and the personnel running it. Having a separate supply for each team ensures that the disruption of one system won't affect everyone.

The initial entry team also needs to take either SABA or SCBA to the victim. Ninety percent of all confined-space incidents are complicated by an atmospheric hazard. It simply makes no sense to work your way toward a victim without being able to supply him with fresh air, given that you find him breathing. This system needs to be attached to the primary team's air supply, since this will allow them to verify that the victim is receiving air.

A team can be placed in the space by a variety of methods. It has been stated before that this is not a text on rope skills, and space limitations preclude doing justice to that subject. As a point of departure, let it be said that you must have good rope skills if you are going to operate in a confined-space environment, equal to or greater than your level of training for confined spaces. For the Operations level in confined spaces, you should have at least the Operations level in rope rescue; for the Technician level in confined spaces, you should likewise be rated at the Technician level in rope. This is in keeping with horse sense, practical experience, and NFPA 1670, in that order.

You should never rely on one system alone to deliver personnel into a space and retrieve them from it again. When placing and removing teams vertically, always strive to establish an overhead anchor system, if possible. This type of arrangement will allow you to place members carefully into the hole; moreover, it will allow you to bring out a victim to a level at which you can work with him easily. Always incorporate a mechanical advantage system into your setup, and back it up further with some form of fall-arresting device or belay line for entry personnel, no matter how short the entry into the space. Rope grabs or mechanical brakes should be a standard component of any lowering or hauling system.

Despite all the merits of tripods, davits, ladder derricks, and existing overhead anchors, many times, in the real world, you'll have to do the best that you can do. If you're in a situation where it's impossible to rig a viable system from an overhead anchor point, you may just have to settle for a 3:1 or a 5:1 system leading out over some edge protection at the hole. You may have to haul the victim up to the opening and then just manhandle him out to safety. This isn't the optimum, most elegant way of performing a rescue, but sometimes circumstances make the use of brute force a necessary evil.

Make sure that you always protect the rope from the edge, and never allow nylon to come in contact with another piece of nylon or a sharp edge. Use standard knots that are strong, easy to tie, and recognized by all members of the team. If your anchors are marginal, always incorporate a load-sharing system and back up the primary anchors. The load placed on each anchor in a multi-point system is a function of the angle between the anchors. As a general rule, try not to create angles of greater than 90 degrees.

In building a mechanical advantage system, remember that, if the rope in the pulley system begins at the load, the system will be odd, meaning that there will be a mechanical advantage of 3:1, 5:1, or so on. If the rope begins at the anchor, the system will be even. Only moveable pulleys add mechanical advantage to a system. By physical law, a stationary pulley, one that is affixed to an anchor, provides only a change of direction for the rope.

A couple of key techniques can help air-supply crews manage the hoses running into the space. Sheathing, otherwise known as the umbilical system, involves putting all of the links to the outside world in a single case, usually a piece of two-inch tubular webbing. This will house the air line, communications line, and a piece of 9 mm cord with a carabiner at the end. Before packing them in the webbing, all of these lines should be bound with wire ties to keep them together. With slack relievers at the couplings, most of the

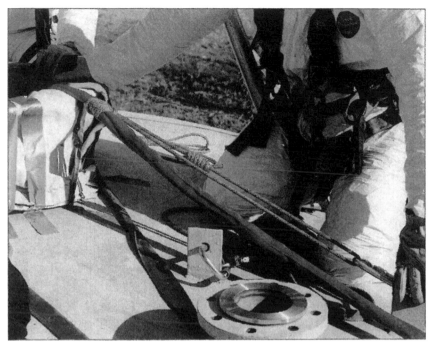

Using a slip knot to manage the deployment of a sheathed umbilical.

weight and tension will fall on the 9 mm cord. When sheathed, these lines can be flaked out on the deck in front of the hole. Laying the lines out in a snakelike fashion will allow easy feeding and takeup. Coils allow for easy deployment, but it isn't as easy to coil a line neatly when taking up. Metal or composite reels, similar to the ones that store garden hose, work very well, but they are heavy and bulky.

Each entrant should have his own air-hose handler. When a team is short of handlers, one person may tend two lines. Consider two to be the maximum for each handler, however. Any more than that, and you'll end up with the lines in a cluster. It is no small matter to a rescue technician on the inside to know that someone is tending his air line appropriately.

In entering a confined space, you are undertaking one of the most dangerous tasks in all of rescue services. In your training, you should have received some tactical and operational parameters

It is no small matter to the technician on the inside to know that someone on the outside is tending his air lines appropriately.

under which you will function in an emergency situation. These guidelines are just that, guidelines, and they are intended to ensure some level of consistency and safety for all members of the entry team. This isn't to imply that individual members should cease using their native intelligence or that they not be allowed to adapt to specific situations that demand improvisation and immediate decisions. Do what it takes to accomplish the mission while ensuring that you will come out again to the surface safely. When deployed as part of a team, think and act as a unit. Work as a team, and communicate your actions to each other. Coordination of effort reduces workload, stress, and time in the hole, and it increases your survivability profile. Maintain adequate communications with the extrication group as well. If the group officer doesn't hear from you on a regular basis, he will rightfully be concerned. Each communications benchmark should include a personal account-ability report, or a statement of your own condition and situation.

From the standpoint of basic survival, it should become almost instinctive that you pay attention to your air line. Team members should assist each other in moving their supply hoses through the space. As you move along, avoid crawling on, lying on, pinching, or chafing these lines. Choreograph your movements and help each other out. Also, constantly assess the location and accessibility of your emergency bypass bottle's on/off valve, and never get yourself in a position where you cannot operate it immediately. Finding out that you can't reach it when you need to is much too late.

If you are in the dark or making many turns, use ingress line to mark the way in and out. Chalk marks, magic marks, or fire line tape can also indicate a path. A member could always follow his air hose to exit a space, but doing so would leave nothing for the next team trying to retrace his route inward.

Always be alert for changes in elevation along the way, as well as unstable footings. Falls are killers in confined spaces. Entrapment is an even greater threat. Although all equipment should have been locked out and tagged out, avoid machinery, electrical equipment, and engulfment hazards at all costs. Only move in and around them when absolutely necessary, and then for the shortest time possible.

Once you locate the victim, you must evaluate your own situ-ation as well as his. If the victim is still breathing, immediately give air to him by means of your SABA. If the victim is not breathing or if his respiration is so compromised that he won't be able to venti-late adequately, begin ventilation with a closed-circuit system. In any situation that you encounter, immediately notify the extrica-tion officer of your findings and actions. He will need to know whether the victim can be extricated quickly by the primary entry team, or whether additional personnel will be required to package the victim and remove him over a long, multiteam, multi-entry

Once you locate the victim, immediately notify the extrication officer of your findings and actions.

operation. Above all, remember what your primary task is, as determined by the extrication officer. Experience shows that most of these events will entail multiple patient removals and possible recoveries.

In support of the primary team, the backup and rapid intervention teams should be directly outside the hot zone in full PPE, ready to go at a moment's notice. They should already have the tools and equipment needed to perform a rescue of the primary entry team or to assist with the removal of victims. As on the fireground, these teams should not be burdened with other tasks. They should be solely dedicated to their mission within the hole in the event that they are called to respond.

If something does go wrong inside the space, the primary team must have an action plan ready to address the emergency. Speaking from experience, the members of the primary team will usually be able to remove themselves from trouble more quickly than exterior personnel can reach them. Although not accounting for all conceivable emergency situations, the following represent the commoner mishaps that might befall personnel inside the hole. Response actions are given for both the interior members and their support crews outside.

LOSS OF AIR

Entry Team Actions

Immediately notify your partner of the air problem. Quickly assess the reason for the loss of air, and ask yourself whether it is a problem that you can correct. Has your partner also lost air, or is the problem only with your system? Pinched air lines, pinched mask lines, or disconnection of the hose at your end are some of the problems that you should be able to correct quickly. Once you have determined the nature of the problem, notify the extrication officer. Begin to use your bypass or emergency escape bottle. If you remain calm, you'll be able to milk the bottle, meaning that you'll only use a little air at a time. Immediately begin to exit the space as a team. If both partners are out of air, the problem is almost certainly on the outside. If only one partner is out of air, monitor his progress and provide buddy breathing as necessary. Indicate to the support crew that you will need removal and that it should be in place by the time you reach the opening.

Exterior Team Actions

Confirm that the entry team has a problem. Notify the air-supply officer and find out whether he knows the cause. Ask him to estimate how long is required for repair. Notify the RIT of the emergency and advise those members to stand by, go on air, and stage. Ascertain from the members inside the hole whether they will need assistance in getting out, as well as their current location. Take appropriate action to restore air to the members of the team and to remove them, either by means of a RIT or by having a retrieval system in place to bring them out quickly. Continue ventilation of the space and enhance it, if possible. Notify the forward medical sector to prepare as necessary. An alternate entry team should prepare to take the place of the first, if appropriate.

INJURY TO AN ENTRY TEAM MEMBER

Entry Team Actions

The injured member should immediately notify his partner that he is injured. Notify the extrication officer as to the extent of the injuries. Indicate whether you can exit as a team or whether assistance is needed. Treat your partner if his injury is life-threatening, and attempt removal, if possible. Await assistance if you are unable to move safely on your own.

Exterior Team Actions

Verify the extent of the injury and the location of the team within the space. Prepare the RIT for entry. Notify the forward medical team and the operations officer. Place teams in the space to assist as necessary. Have a second team ready to continue the operation once the primary team has been removed.

ATMOSPHERIC MONITOR ALARM

Entry Team Actions

Determine action guidelines if the alarm goes off. If it is a flammability alarm, identify a buffer. If it is an oxygen or toxicity alarm, determine the need to exit. Notify the extrication officer of the alarm and advise him of the current conditions in the space. Find out the potential cause of the alarm, determine whether to stay or go, and advise the extrication officer of your decision. Follow whatever orders you receive from the extrication officer.

Exterior Team Actions

If the alarm sounds at the entry point, advise the members on the inside and confirm their interior readings. Advise them of the conditions at the entry point, then ask them whether they feel they can stay. The extrication officer must also be part of this decision. He has the power to order them out of the hole, if necessary. Assess the ventilation efforts and take the necessary steps to improve them.

EXPLOSION AND FIRE IN THE SPACE

Entry Team Actions

If you are able, advise the extrication officer of your status and attempt an immediate escape from the space. Your partner must go with you. In the case of injury, advise the support crew of your location and request immediate assistance. Maintain the firefighter-down position and recheck the integrity of your PPE, breathing apparatus, and other gear. When possible, check on the status of the original victim. Maintain constant communications with the extrication officer.

Exterior Team Actions

Immediately extinguish any remaining fire. Ventilate the space, and attempt to locate the source of the fire or explosion. Immediately contact the entry team and determine the status of those personnel. If necessary, place the RIT in the space on a search-and-rescue mission. If possible, flood the space with foam to reduce the chance of a secondary blast. Bring forward medical personnel to the hot zone, and advise the operations officer of the situation. Treat those personnel on the outside who may be injured. Prepare additional teams for entry and rescue as needed. Have an extrication team ready to remove the injured. Request the necessary medical support through operations.

165

CHAPTER QUESTIONS

1. True or false: According to OSHA's two-in, two-out rule for imminently IDLH environments, you must always have two additional personnel outside the space for each two inside.

2. True or false: Nylon rope is best protected if it is allowed to pass over another piece of nylon.

3. If the rope in the pulley system begins at the load, the system will be _____.

4. What is sheathing?

5. An air-line handler should be responsible for managing no more than how many lines?

6. Why should an entry technician not rely on his air line as a marker for his path of egress?

7. True or false: Despite the value of a RIT, the members of the primary team will usually be able to remove themselves from trouble more quickly than exterior personnel can reach them.

8. True or false: If the entry team experiences a loss of air, the RIT should deploy immediately and automatically.

CHAPTER TEN

REMOVING THE VICTIM

All of the effort described up to this point is aimed at reaching and removing a victim from a potential tomb. Invariably, this rather straightforward assignment is made infinitely more complex by the hostile nature of the spaces in which people become trapped. In many ways, all of the necessary attention given to environmental concerns takes much of the immediate focus of responders away from the victim, shifting it instead toward an array of logistical concerns centered on the physical act of getting into the hole and safely getting out again. Never to be forgotten in the rescue equation, however, is the manner of medical care that you can provide to the victim once you locate him. By the very nature of these operations, the caliber of the medical component is of prime concern to the rescuer as well, for the member who ventures into the hole is never more than one mischance away from becoming a victim himself.

If the victim can easily be removed by the primary team by means of a hasty hitch or wristlets, then the removal phase of the operation may quickly be brought to a conclusion. In most cases, however, the process will require much more time and planning. It may require multiple entries and a high degree of coordination between the interior and exterior teams. Regardless of the speed or complexity of the operation, two basic task assignments apply to every removal and recovery operation. In essence, the interior team is responsible for locating, disentangling, packaging, and moving the victim back to the entry point, whether by their own means or with the help of additional rigging from outside. The exterior team is responsible for rigging any system required to remove the victim from the space.

The exterior team is responsible for rigging any system required to remove the victim from the space.

When packaging a patient, you must consider the extent of his injuries. Obviously, the type of transportation device that you choose depends on the shape and layout of the space in which you're working. When moving a victim in a confined space, always try to position yourself on the egress side of him; that is to say, stay between the victim and the opening. Avoid being blocked in by the victim when moving through pinch points. This may not always be possible due to the physical layout of the space, the size of the victim, and the limited leverage that you have on him. If, for some reason, you must move the patient in such a way that it will place him between you and the exit, do so quickly and smoothly. Plan the move and communicate your intentions with your partner. Keep the egress point blocked for as little time as possible. If you find that you cannot move the patient through tight quarters, stop, back up, and regroup. Make the exterior teams aware of any difficulty that you're having. You might need a team to enter from the opposite side to assist you in moving the victim through the tight quarters. Anticipate the movement of the air hoses in advance. Ensure that these lifelines are clear of the pinch points prior to making the move. Manage them in such a way that they don't entangle on the space or around the victim. If you are working in a potentially flammable atmosphere, try to minimize the dropping and scraping of hardware. Since you should be operating in an environment at a fraction of the LEL, this may not be a critical issue; however, it is always better to be as safe as possible. As always, exercise care if you must use wristlets on burned limbs, since these can pull the skin right off in some circumstances.

If multiple teams are required to perform multiple entries, the extrication officer should ensure that each outgoing crew directly exchange information with the ingoing crew. This briefing should encompass not only the present location of the victim and his status, but also whether the exiting team's task assignment was fully accomplished or not. Naturally, any specific hazards, including decon concerns, should be mentioned. In describing the situation as accurately as possible, it is best if the exiting members provide their update using a map, drawing, or blueprints of the space.

During the recovery and removal phase, there is a tendency for exterior teams to become complacent, since they aren't yet directly involved in handling the victim. Should an emergency arise, or when the entry team and victim are ready for removal, the exterior team must be ready to go to work immediately. For that reason, specific tasks should be given to exterior teams during active entry operations. Line tending, atmospheric monitoring, communications, and maintenance of the air supply are just a few of the more obvious assignments to be made. All of the systems put in place must be in accordance with the victim's size and weight. Once the actual removal is underway, the haul team needs to maintain as much C-spine control as possible. Mechanical advantage systems are much preferred over brute force; however, you should never use electrical winches or other nonmanual mechanical means, since your inability to judge resistance may result in severe injuries to the victim. Dislocation, dismemberment, and death are all possibilities if you should continue to exert force after a limb or other body part has snagged. In preventing this, the team should decide early on whether the patient is to be removed headfirst, feetfirst, or supine.

Once the victim has been removed from the space, transfer him immediately to the forward medical team. Lifesaving care at the BLS and ALS levels must begin at once. Since the patient may need decontamination, the receiving medical team should be protected with PPE. Patient care is the responsibility of the forward medical team and the decontamination team. Under no circumstances should the extrication team get caught up in or committed to this process.

Medical provision at the scene of a confined-space incident involves the delivery of care to both victims and team members, and the protocols are based entirely on the hazards to which patients have been exposed. The administration of frontline medicine is sometimes critical to the success of the rescue. In the event of an actual rescue operation, there will likely be more than

Once the victim has been removed from the space, transfer him immediately to the forward medical team.

one patient. Even when there is only one victim, there is always a high chance, in both rescue and recovery scenarios, that medical care will be required by team members within the space. Let it be said that there are very few confined-space incidents of medium length. If the victim can't be extracted rather quickly, then the operation will tend to become either a painstaking rescue or a lengthy recovery. Appropriately trained medical personnel may find themselves in a space for hours on end, providing care throughout ongoing extrication operations. Frontline medicine, when it is required, is most often a component of incidents involving structural collapse, trench collapse, and in some confined spaces where atmospheric problems aren't the chief concern.

For those scenarios in which the atmosphere has been compromised, rapid extrication with no up-front medicine will provide the best potential outcome. Once removed from the space, you can expect victims to present an array of respiratory problems, ranging from apnea to dyspnea, and neurological states, ranging from unconsciousness to combative behaviors. In some instances, you will find yourself dealing with a patient lapsing into cardiac arrest. Still, most of the victims exposed to significant atmospheric problems will have been declared dead long before extrication teams can remove them from the space.

Unique to the delivery of frontline medicine in a structural collapse, possibly any confined space, is the need for medical personnel to amputate limbs, or even section the body of a corpse, often to clear a path to a live victim. Medical team members, in consultation with rescue members, must decide whether other methods exist that will allow them to reach viable patients first, since sectioning can have a psychological impact on all who witness it, and it creates a biological hazard as well. In truth, given modern technology and techniques, there are very few situations in which sectioning should be necessary. Usually, when you read about such an event, you will find that the sectioning of limbs during a confined-space operation has been done by medical personnel thrust into that environment with little or no training. Often the rescue team supporting them is similarly undertrained. In these situations, the arm or leg is amputated to expedite extrication, either because the team members believed that they were in extreme danger or because they felt they couldn't perform the necessary extrication to liberate the trapped limb. Either judgment likely indicates a shortfall in tools or training.

This isn't to imply that sectioning is never an option. Still, by delivering frontline medicine, you can probably buy the time that rescue workers need to extricate the victim properly.

In a trench environment, it isn't unusual to find that the chest has been covered, threatening asphyxiation. When this occurs, you must immediately clear the chest area so that it can expand with the normal cycles of respiration. Often this is difficult unless you can rapidly shore up the surrounding areas to forestall a secondary collapse. Femur and pelvic fractures are common in incidents where plate steel, concrete pipe, or some other heavy object has pinned a victim without killing him outright. Crush death syndrome is frequently a consideration in such instances. Less common is impalement by pickets, rebar, and other stakelike objects. Certainly the medical team should be able to deal with all of these circumstances, although they don't all rank in the same order of likelihood.

Tight spaces, dusty environments, compromised atmospheres, the restrictions of PPE, exposure to toxins, contact with body fluids, the threat of secondary collapse, sudden claustrophobia, the pressures to perform—all of these can conspire against even the most physically fit rescuer. Fortunately, in the vast majority of cases, all the care that is required for the team can be found in any good rehab section, in the form of hydration, some monitoring of baseline vitals, and perhaps some limited decon. Still, medical team members need to be prepared to treat any number of occupational injuries, including significant dehydration, strains, sprains, back injuries, smashed fingers, and the like.

In short, the members of the medical team must understand the parameters under which they will work and be prepared to treat everything from the ordinary to the extraordinary. Although they may be fine medical personnel, street medics without special training do not belong in a technical rescue environment. Any member of the organic medical component (i.e., those medics who are specially trained and equipped to go wherever the rescue team goes) must receive the same level of training as the rescue technicians. Although their primary task isn't the same, they need to be able to confront the same environments. Given the staffing cutbacks in modern municipal departments, it wouldn't be unusual for a member of the medical team to be called on to function as a member of the entry team, extrication team, removal team, or support team. Each of these assignments require an intimate knowledge of the related equipment, systems, and procedures. For this reason, every member of the medical team should have, at minimum, Operations-level training in rope rescue, confined-space rescue, structural collapse rescue, trench rescue, and haz mat operations. Training for helicopter operations or swift-water rescue may also be necessary, depending on the geographical context.

As stated above, most confined-space incidents involve more than one victim. Responders should not be surprised to see multiple body bags or patients. According to OSHA, about 35 percent of victims at confined-space incidents are would-be rescuers. This means that one in three victims at an industrial or municipal site will not be the original entrant. In some instances, these personnel are entrapped and overcome by the same environment that took the original victim. In other instances, rescuers who attempt to enter the space may be driven off by contaminated atmospheres or physical obstacles, and the injuries they sustain may hinge on the length of time that they were exposed to the hazard. Of the thirty or so actual confined-space operations in which I have participated, all but one involved more than one patient or fatality. The sobering reality is that, in industrial and municipal spaces, the vast majority of them are fatalities.

In a trench environment, any of several scenarios may play out. It isn't unusual to find one victim buried underneath another. Of course, the lower victim often fares much worse. Usually the lower victim will be dead, whereas the upper victim will be suffering from pelvic injuries or large-bone fractures from the impact of tons of earth. Extrication times are often long, since hand movement of the soil is often necessary to remove the top patient. Additionally, he may be in a kneeling position, with his legs bent at the knees and a sheet of plywood or some other object solidly locked down on his calves, along with several cubic yards of soil. You may also find that workers initially in the trench scrambled out during the collapse, so don't neglect looking for ambulatory patients who might have sustained fractures or other impact injuries.

Structural collapse certainly offers a unique challenge for medical personnel. Depending on the type of structure, occupancy, time of day, and the cause of the collapse, you may find yourself faced with a wide variety of scenarios, all in the same incident. Single victims may be trapped near the surface, and dozens more may be entombed. Medical teams must be prepared to treat any number of patients, from the lightly wounded to those who require hours and possibly days to extricate.

The Oklahoma City bombing, although an atypical event, is certainly a good example of what the range of medical response might be. Very little medical care and delivery by technical-rescue medical teams was provided to victims trapped or injured in the

Although an atypical event, the Oklahoma City bombing provides
a good example of what the range of medical response might be.

explosion and collapse. Initial care for the lightly wounded and the lightly trapped was provided by first-response personnel, and their removal to a medical facility was accomplished quickly. Still, technical-rescue medics and physicians had a tremendous role to play in terms of supporting the members performing recoveries. This aspect of care is often overlooked by organizations and individual members who falsely believe that only the original victim suffers the burden of risk.

One confined-space event to which we responded involved a triple fatality on the Nimitz-class carrier USS Truman. While under construction in the yard, three workers were killed in a small working space when they disconnected a waste line that was still active. Raw sewage flooded into the space and immediately contaminated the interior atmosphere. The reconnaissance and recovery operations took place four levels below the hangar deck. Although no viable victims were brought out, the fourteen-hour recovery operation required considerable medical support for the team members.

Medical treatment and decontamination should follow local protocols. The following suggestions have proved quite successful over the years.

1. Give immediate attention to airway, breathing, and circulation problems. Intubation and supplemental capture of the airway may be important. Combative, hypoxic patients should be paralyzed with an appropriate drug and their airway captured.

2. Consider treatment for hypothermia, per the local protocol.

3. Implement a fluid challenge as necessary using lactated ringers or saline.

4. Provide additional cervical spine support as necessary.

5. If there has been exposure to a hazardous material, take the appropriate steps immediately, even before extrication, to identify that substance and to specify a treatment for it.

Similarly, decon procedures should be based on both the substance involved and your local haz mat protocols. There are five general-purpose decon solutions for the various classes of hazardous materials that you will encounter. The first three may find application, as appropriate, when dealing with an unknown material.

Decon Solution A: A solution containing 5 percent sodium carbonate and 5 percent trisodium phosphate. Mix four pounds of commercial-grade trisodium phosphate with each ten gallons of water. These chemicals are available in most hardware stores. Sodium carbonate is also known as soda ash, and trisodium phosphate is common in some laundry detergents. This solution is a base and a surfactant.

Decon Solution B: A solution containing 10 percent calcium hypochlorite. Mix eight pounds with each ten gallons of water. Calcium hypochlorite is commonly called HTH and is available from swimming pool supply stores. This solution is acidic.

Decon Solution C: A general-purpose rinse for both of the above solutions is a 5-percent solution of trisodium phosphate. To prepare this, mix four pounds with each ten gallons of water. This solution is a surfactant.

Decon Solution D: A diluted solution of hydrochloric acid. Mix one pint of concentrated HCl into ten gallons of water. Stir with a wooden or plastic stirrer. This solution is acidic.

Decon Solution E: A concentrated solution trisodium phosphate, mixed into a paste and scrubbed with a brush. Rinse with water.

The following is offered as a guideline.

1. Inorganic acids, metal-processing wastes: Solution A.

2. Heavy metals, such as mercury, lead, cadmium, and the like: Solution B.

3. Pesticides, chlorinated phenols, dioxins, and PCBs: Solution B.

4. Cyanides, ammonia, and other nonacidic inorganics wastes: Solution B.

5. Solvents and organic compounds such as trichloroethylene or PBBs: Solution C or A.

6. Oily, greasy, unspecified wastes not suspected of containing pesticides: Solution C.

7. Inorganic bases, alkali, or caustic waste: Solution D.

8. Radioactive materials: Solution E.

9. Etiologic materials: Solution A or B.

Despite these guidelines, always seek expert assistance in deciding which solution to use. The manufacturer, MSDS, poison control centers, medical specialists, and research materials are some of your options. An emergency decon is always the most efficient method of application. Use copious amounts of water and a decon solution. Nine-step decons aren't practical or time-effective for confined-space operations. Remember, too, that you must set up decon stations for nonambulatory victims. Unconscious or packaged patients need to be processed in the same manner as any other.

CHAPTER QUESTIONS

1. When moving a victim in a confined space, where should you always try to position yourself?

2. True or false: The extrication officer should ensure that each outgoing crew directly exchange information with the ingoing crew.

3. Why should specific tasks be given to exterior teams during active entry operations?

4. Why should you never use electrical winches or other nonmanual mechanical means to raise a victim out of a confined space?

5. True or false: The extrication team should assist the forward medical team and decon personnel in handling the patient.

6. True or false: In scenarios in which the atmosphere has been compromised, rapid extrication with no up-front medicine will provide the best potential outcome.

7. True or false: According to the author, every member of the medical team should have, at minimum, Operations-level training in rope rescue, confined-space rescue, structural collapse rescue, trench rescue, and haz mat operations.

8. True or false: Most confined-space incidents involve only one victim.

CHAPTER ELEVEN

TERMINATION ACTIVITIES

E ven after all of the team members have crawled out of the hole at the conclusion of an incident, a number of responsibilities still remain. Termination is actually the beginning of a new phase. If conducted properly, it will provide you with a wealth of information. It is during this phase that you can document the incident thoroughly, not only for the learning experience, but also to protect yourself and the organization against any inquiry, investigation, or litigation that may follow. The operations officer is the entity who should collect all of the tactical work sheets, notes, blueprints, maps, and any other documentation of the incident. All of this information should be collected to produce a postincident report, as well as to create a central file on the operation.

Of foremost importance, a coordinated termination effort is necessary to account for all personnel and victims. Equipment may also be counted at this time. You should demobilize the units according to a plan developed through the incident management system and codified in your SOPs. Secure the accident site, and ensure that it is released to the proper authorities. All of the documentation that has been amassed during the course of the incident should now be collected and stored in one file. Already, in the wake of the incident, you will be able to anticipate the nature of your postincident briefing.

Use your personnel accountability system to verify the whereabouts and safety of all of your personnel. Make sure that the tactical worksheet has been updated, including exit times and off-air times. Medical screening and rehab information should also be updated. Send all entry team personnel to rehab, where the medical team can capture a full set of vital signs, as well as any other significant information. As members exit the site, do not let them forget to remove any lockout, tag-out devices that they have placed during the incident.

From the data and diagrams that you collect, you should now begin to document in full the course of your operations. This process may begin with an accurate diagram, detailing the location and positions of the victims within the space. Illustrate and describe in words also the surroundings in which they were found. Were they without breathing apparatus upon discovery? Was that apparatus full or empty? Was there evidence of a fall? Trauma? Burns? Were any mechanical systems connected with the space operational when responding units arrived? Documenting the details now will put you in a better standing should you have to answer to OSHA or some other investigative inquiry at a later date. A prime aspect of this process is to compile a list of witnesses and coworkers at the site during the incident. The input of the entry team is critical, since it is they who can provide the best information for creating accurate maps or drawings of the space. Include statements or any other documentation of any problems that were encountered in the space, including environmental hazards and operational difficulties. If you have a required OSHA document on hand, complete it at this time, while the incident is still fresh in everyone's mind. If it hasn't be done already, contact the local OSHA office or the state OSHA representative. Naturally, if the site has been declared a crime scene, you should coordinate your documentation efforts with the appropriate law enforcement agency.

A plan may be required to demobilize a complex incident in an orderly fashion.

Depending on the size of the incident, the number of personnel involved, the geographical boundaries of the operation, and the length of time that units have been committed, you will likely need a demobilization plan. In some instances, that plan may be very straightforward, involving only a few steps. Other times, there may be thirty vehicles and a hundred personnel at the scene, and a full plan is absolutely essential. Even a one-page planning document can help maintain an orderly takeup and demobilization of the units. Time is an important factor in any demobilization, and any officer who oversees this phase must be aware of how long it can take to inventory equipment. Doing so can be further complicated if that equipment needs to be decontaminated or discarded. Each officer should be able to estimate how many members are required to complete their portion of the task. Release from the site isn't always a matter of who finishes first, since certain personnel are critical to the postoperation and must remain. In the case of a multiagency response, it may help to break outside units first, replacing those members with home personnel. Demobilization is an operation in and of itself, and the overseeing officer must have a grasp of what internal or external support is available. In terms of preparedness, it is important to know how quickly demobilized specialty units can be back in service and ready to respond to the next incident.

Whenever you turn over a site to another agency, you should pass along certain information. Document the transfer of control. Be sure to get the name, address, and title of the responsible party to whom you are releasing the site. Also get a phone number in case you need to contact that person later, perhaps to recoup costs for equipment lost or damaged during the operation. Note the exact time that you turned over the site, and document that you provided the responsible party with any safety information regarding how the site was left or precautions to be taken until specific problems have been corrected. Provide a list of the lockout, tag-out systems that were removed. Before leaving, make sure that access to the space is restricted. If it isn't closed off, at least post a warning or stretch fire line tape around the general area.

Always police the grounds after an event. Clean up any refuse that your units may have generated during the operation. This may sound trivial, considering what you have just been through, but there may be repercussions later if the owner or responsible party begins to squawk about how your teams trashed his place of business. If you had to break or destroy any property during the course of the incident, give a list of those items to the responsible party, and keep a copy on file.

I have never been to a confined-space operation in which equipment was not lost, damaged, left in the space, or placed on someone else's rig. The logistics officer should document all lost and damaged equipment. This will help you place damaged equipment out of service and get it repaired as quickly as possible. Since specialty equipment is at a premium, the turnaround time on repairs should be as short as possible. Documenting equipment casualties will also help you evaluate whether your systems and devices are holding up under the rigors of real-world operations. There is no better proving ground than the real world, and if you continually have problems with a certain implement, connection, or apparatus, perhaps there is either a flaw in the design or it is being used improperly. Keeping track of equipment this way also serves the maintenance program. If you have a cost-recovery program, careful documentation will help you keep track, for billing purposes, of what is broken, lost, or in need of repair. Many items don't get lost, but instead wander onto other rescue trucks or into the equipment bags of other team members, and this, of course, needs to be controlled. The decontamination phase is one juncture at which equipment may readily be returned to the rightful owner.

Reconstructing the incident by means of a postincident analysis is an important way to assess, not only the overall chain of events, but also the effectiveness of the members and their methods. The main purpose of a postincident analysis is to reinforce actions and procedures that are effective and to give the command staff insight into how the organization can better itself. A comprehensive analysis can foster improvements in everything from response times and training needs to staff requirements and working relationships among the responders. This sort of formal postincident analysis is initiated by the incident commander and scheduled with the appropriate response personnel and assisting agencies.

This type of analysis should be conducted after every confined-space operation. The incident commander and all other division officers who were responsible for tactical decisions should complete a fact sheet as soon as possible after the incident and submit it to the IC. A session is then scheduled, and all of the participants are notified once all of the fact sheets have been completed. The incident commander normally leads the proceedings and assigns someone to record the minutes of it. The discussion should be open and honest regarding all aspects of the operation. It should be noted that the fact sheets and summary of the analysis are not public documents. The work sheets are internal documents used for evaluation and training. They should not be attached to any formal report, nor should reference be made to them in any public context.

Done correctly, a timely postincident analysis will provide you with a wealth of information. It allows each member of the command team and the operational members to see what went on outside the sphere of their own influence. Understanding the big picture will foster cooperation to some degree. It will also alert them to any flaws or deficiencies in their methods of operating.

There is a huge psychological component to any rescue or recovery activity in a confined space. Still, the need for a critical incident stress debriefing (CISD) after such an event is a debatable issue. I will confess to you that I have never been a big fan of CISD. There are certainly times that we all need to talk to our peers about the pressures of a given event. Still, I have never found any confirmation in the medical literature indicating that a stress debriefing adds to either health or happiness. Any fledgling responder already knows full well that the demands of his chosen career are high, and certainly my comments aren't meant to imply that I am insensitive to those who may be emotionally troubled by what they encounter at an emergency scene. You may find it necessary to bring in your local CISD team and provide confidential sessions for those personnel who request them. There should be a continual opportunity for members to get whatever assistance they feel is necessary to allow them to cope with the events to which they have been exposed. But it shouldn't be mandatory for all personnel.

As I write this, I am thinking of my last night in Oklahoma City. The operation had shut down, and two other rescuers and I were each given a piece of paper and told to report to the CISD building. All of us indicated that we were fine and had no desire to go through any form of pseudotherapy at that point. Naturally, this incident had affected each of us in different ways—we had just experienced a life-changing event, but we were far from packing it all in out of emotional devastation.

Once in the office, the debriefer greeted us, asking how we felt and suggesting that we all bare our souls about our long days working under the pile. Our collective reply to him, with no trace of a smile from anyone, was "Sign the papers or we're going to kill you." The debriefer duly gave his signature to each of us. Having successfully obtained our note from the teacher, we left, each of us departing to our home and whatever brand of comfort awaited us there.

CHAPTER QUESTIONS

1. Who should collect all of the documentation at the end of an incident?

2. How should units be demobilized?

3. The best way to verify the whereabouts and safety of all of your personnel is to use a _____.

4. What government agency must you contact in the event of a confined-space incident?

5. Why isn't release from a site a matter of who finishes first?

6. Describe some of the basic steps involved in releasing an emergency site to the responsible party.

7. Why should you document damaged equipment?

8. Who normally initiates and leads a posincident analysis?

APPENDIX

Evolution of Confined-Space Standards

The confined-space standard for general industry has come about in stages, through a process of evolution. OSHA, NIOSH, and a variety of state organizations have been developing and printing documents on confined-space safety, operations, and problems since the mid-1970s. Below is a time line of some of the highlights of this process.

July 1975: OSHA issues *Advanced Notice of Proposed Rules Making (ANPR), Standard for Work in Confined Spaces*. The purpose is to obtain information that might be useful in the development of a standard.

December 1979: NIOSH issues *Criteria for a Recommended Standard: Working in Confined Spaces*. In this document is a recommendation for the training of employees and the need for a permit program. The document is sixty-three pages long, and it outlines a variety of familiar themes.

January 1986: NIOSH publishes an alert, *Request for Assistance in Preventing Occupational Fatalities in Confined Spaces*. The report focuses on case studies involving circumstances leading to the death of sixteen workers. It notes that "more than 60 percent of fatalities in confined spaces occur to would-be rescuers." This figure has dropped to 35 percent since the release of 1910.146.

February 1987: OSHA revises the *Construction Industry Safety and Health Standards*, 29 CFR 1926/1910, which includes Section 1926.417 on the lockout and tagging of circuits, and Subpart S, Sections 1926.800 to 1926.802, outlining safety precautions to be taken in tunnels, shafts, caissons, and cofferdams.

July 1987: NIOSH publishes a pamphlet, *A Guide to Safety in Confined Spaces,* which outlines specific steps toward preventing injuries and death in confined spaces, and even includes checklists for this process.

June 1989: OSHA issues a *Notice of Proposed Rule Making.* This outlines the requirements of the proposed standard and solicits comments from the regulated community.

September 1989: OSHA promulgates *The Control of Hazardous Energy Sources (Lockout/Tag Out),* 29 CFR 1910.147 (Appendix 2).

November 1990: The comment periods are finally closed.

January 1991 to January 1993: OSHA generates a list of all confined-space fatalities and injuries for a two-year period, outlining the causes, the employers, the outcomes, and the fines levied.

January 1993: OSHA publishes the *Final Rule* in the Federal Registrar, more than seventeen years after the first *ANPR* is issued.

June 1993: OSHA publishes corrections to the *Final Rule* in the Federal Registrar.

January 1994: NIOSH publishes *Worker Deaths in Confined Spaces: A Summary of Surveillance Findings and Investigative Reports.* This document outlines a wide variety of deaths and injuries in confined spaces, taking an in-depth look at the causes and corrective matters.

August 1994: OSHA holds hearings at its Washington office at the request of the United Steelworkers of America. In attendance to provide testimony are a variety of concerned parties, both public and private, including IAFC, for whom the author testifies. Suggested changes to the standard are placed in official hearing records for use in formulating updates.

The following is a reprint of Appendix F of OSHA 1910.146.

OSHA 1910.146
Appendix F

- **Standard Number:** 1910.146 APP F
- **Standard Title:** Permit-required confined spaces
- **SubPart Number:** J
- **SubPart Title:** Non-Mandatory Appendix F—
  Rescue Team or Rescue Service Evaluation Criteria
- **Applicable Standard:** Applicable Standard:

Non-Mandatory Appendix F—
Rescue Team or Rescue Service Evaluation Criteria

(1) This appendix provides guidance to employers in choosing an appropriate rescue service. It contains criteria that may be used to evaluate the capabilities both of prospective and current rescue teams. Before a rescue team can be trained or chosen, however, a satisfactory permit program, including an analysis of all permit-required confined spaces to identify all potential hazards in those spaces, must be completed. OSHA believes the compliance with all the provisions of §1910.146 will enable employers to conduct permit space operations without recourse to rescue services in nearly all cases. However, experience indicates that circumstances will arise where entrants will need to be rescued from permit spaces. It is therefore important for employ-

ers to select rescue services or teams, either on-site or off-site, that are equipped and capable of minimizing harm to both entrants and rescuers if the need arises.

(2) For all rescue teams or services, the employer's evaluation should consist of two components: an initial evaluation, in which employers decide whether a potential rescue service or team is adequately trained and equipped to perform permit space rescues of the kind needed at the facility and whether such rescuers can respond in a timely manner, and a performance evaluation, in which employers measure the performance of the team or service during an actual or practice rescue. For example, based on the initial evaluation, an employer may determine that maintaining an on-site rescue team will be more expensive than obtaining the services of an off-site team, without being significantly more effective, and decide to hire a rescue service. During a performance evaluation, the employer could decide, after observing the rescue service perform a practice rescue, that the service's training or preparedness was not adequate to effect a timely or effective rescue at his or her facility and decide to select another rescue service, or to form an internal rescue team.

A. Initial Evaluation

I. The employer should meet with the prospective rescue service to facilitate the evaluations required by §1910.146(k)(1)(i) and §1910.146(k)(1)(ii). At a minimum, if an off-site rescue service is being considered, the employer must contact the service to plan and coordinate the evaluations required by the standard. Merely posting the service's number or planning to rely on the 911 emergency phone number to obtain these services at the time of a permit space emergency would not comply with paragraph (k)(1) of the standard.

II. The capabilities required of a rescue service vary with the type of permit spaces from which rescue may be necessary and the hazards likely to be encountered in those spaces. Answering the questions below will assist employers in determining whether the rescue service

194

is capable of performing rescues in the permit spaces present at the employer's workplace.

1. What are the needs of the employer with regard to response time (time for the rescue service to receive notification, arrive at the scene and set up and be ready for entry)? For example, if entry is to be made into an IDLH atmosphere, or into a space that can quickly develop an IDLH atmosphere (if ventilation fails or for other reasons), the rescue team or service would need to be standing by at the permit space. On the other hand, if the danger to entrants is restricted to mechanical hazards that would cause injuries (e.g., broken bones, abrasions) a response time of 10 or 15 minutes might be adequate.

2. How quickly can the rescue team or service get from its location to the permit spaces from which rescue may be necessary? Relevant factors to consider would include: the location of the rescue team or service relative to the employer's workplace, the quality of roads and highways to be traveled, potential bottlenecks or traffic congestion that might be encountered in transit, the reliability of the rescuer's vehicles, and the training and skill of its drivers.

3. What is the availability of the rescue service? Is it unavailable at certain times of the day or in certain situations? What is the likelihood that key personnel of the rescue service might be unavailable at times? If the rescue service becomes unavailable while an entry is underway, does it have the capability of notifying the employer so that the employer can instruct the attendant to abort the entry immediately?

4. Does the rescue service meet all the requirements of paragraph (k)(2) of the standard? If not, has it developed a plan that will enable it to meet those requirements in the future? If so, how soon can the plan be implemented?

5. For off-site services, is the service willing to perform rescues at the employer's workplace? (An employer may not rely on a rescuer who declines, for whatever reason, to provide rescue services.)

6. Is an adequate method for communications between the attendant, employer and prospective rescuer available so that a rescue request can be transmitted to the rescuer without delay? How soon after notification can a prospective rescuer dispatch a rescue team to the entry site?

7. For rescues into spaces that may pose significant atmospheric hazards and from which rescue entry, patient packaging and retrieval cannot be safely accomplished in a relatively short time (15-20 minutes), employers should consider using airline respirators (with escape bottles) for the rescuers and to supply rescue air to the patient. If the employer decides to use SCBA, does the prospective rescue service have an ample supply of replacement cylinders and procedures for rescuers to enter and exit (or be retrieved) well within the SCBA's air supply limits?

8. If the space has a vertical entry over 5 feet in depth, can the prospective rescue service properly perform entry rescues? Does the service have the technical knowledge and equipment to perform rope work or elevated rescue, if needed?

9. Does the rescue service have the necessary skills in medical evaluation, patient packaging and emergency response?

10. Does the rescue service have the necessary equipment to perform rescues, or must the equipment be provided by the employer or another source?

B. Performance Evaluation

Rescue services are required by paragraph (k)(2)(iv) of the standard to practice rescues at least once every 12 months, provided that the team or service has not successfully performed a permit space rescue within that time. As part of each practice session, the service should perform a critique of the practice rescue, or have another qualified party perform the critique, so that deficiencies in procedures, equipment, training, or number of personnel can be identified and corrected. The results of the critique, and the corrections made to respond to the deficiencies identified, should be given to the employer to enable it to determine whether the rescue service can quickly be upgraded to meet the employer's rescue needs or whether another service must be selected. The following questions will assist employers and rescue teams and services evaluate their performance.

1. Have all members of the service been trained as permit space entrants, at a minimum, including training in the potential hazards of all permit spaces, or of representative permit spaces, from which rescue may be needed? Can team members recognize the signs, symptoms, and consequences of exposure to any hazardous atmospheres that may be present in those permit spaces?

2. Is every team member provided with, and properly trained in, the use and need for PPE, such as SCBA or fall arrest equipment, which may be required to perform permit space rescues in the facility? Is every team member properly trained to perform his or her functions and make rescues, and to use any rescue equipment, such as ropes and backboards, that may be needed in a rescue attempt?

3. Are team members trained in the first aid and medical skills needed to treat victims overcome or injured by the types of hazards that may be encountered in the permit spaces at the facility?

4. Do all team members perform their functions safely and efficiently? Do rescue service personnel focus on their own safety before considering the safety of the victim?

5. If necessary, can the rescue service properly test the atmosphere to determine if it is IDLH?

6. Can the rescue personnel identify information pertinent to the rescue from entry permits, hot work permits, and MSDSs?

7. Has the rescue service been informed of any hazards to personnel that may arise from outside the space, such as those that may be caused by future work near the space?

8. If necessary, can the rescue service properly package and retrieve victims from a permit space that has a limited size opening (less than 24 inches (60.9 cm) in diameter), limited internal space, or internal obstacles or hazards?

9. If necessary, can the rescue service safely perform an elevated (high angle) rescue?

10. Does the rescue service have a plan for each of the kinds of permit space rescue operations at the facility? Is the plan adequate for all types of rescue operations that may be needed at the facility? Teams may practice in representative spaces, or in spaces that are "worst-case" or most restrictive with respect to internal configuration, elevation, and portal size. The following characteristics of a practice space should be considered when deciding whether a space is truly representative of an actual permit space:

(1) Internal configuration.

(a) Open—there are no obstacles, barriers, or obstructions within the space. One example is a water tank.

(b) Obstructed—the permit space contains some type of obstruction that a rescuer would need to maneuver around. An example would be a baffle or mixing blade. Large equipment, such as a ladder or scaffold, brought into a space for work purposes would be considered an obstruction if the positioning or size of the equipment would make rescue more difficult.

(2) Elevation.

(a) Elevated—a permit space where the entrance portal or opening is above grade by 4 feet or more. This type of space usually requires knowledge of high angle rescue procedures because of the difficulty in packaging and transporting a patient to the ground form the portal.

(b) Non-elevated—a permit space with the entrance portal located less than 4 feet above grade. This type of space will allow the rescue team to transport an injured employee normally.

(3) Portal size.

(a) Restricted—A portal of 24 inches or less in the least dimension. Portals of this size are too small to allow a rescuer to simply enter the space while using SCBA. The portal size is also too small to allow normal spinal immobilization of an injured employee.

(b) Unrestricted—A portal of greater than 24 inches in the least dimension. These portals allow relatively free movement into and out of the permit space.

(4) Space access.

(a) Horizontal—The portal is located on the side of the permit space. Use of retrieval lines could be difficult.

(b) Vertical—The portal is located on the top of the permit space, so that rescuers must climb down, or the bottom of the permit space, so that rescuers must climb up to enter the space. Vertical portals may require knowledge of rope techniques, or special patient packaging to safely retrieve a downed entrant.

[63 FR 66039, Dec. 1, 1998]

SAMPLE CONFINED-SPACE OPERATING GUIDELINE

CONFINED-SPACE ENTRY AND RESCUE

Purpose

To establish procedures for entry and rescue operations in a confined space.

Scope

These procedures apply to all uniformed personnel.

Phase I: Scene Preparation

On arrival at a confined space, the first-due technical rescue team personnel should obtain the following information from the first-due company officer, battalion officer, or job-site foreman. The (hometown) Fire Department shall assume command and control of any incident involving a confined-space entry and rescue that occurs within its jurisdiction.

Step One: Assessment

1. What type of space is this?

2. Are there product-storage hazards?

3. Locate and secure the job site foreman or a reliable witness.

4. Determine the location and number of victims.

5. Obtain blueprints or maps, or have on-site personnel draw a sketch of the site.

6. Determine the mechanisms of entrapment or the nature of illness.

7. Make a conscious decision as to whether this is a rescue or a recovery.

8. Determine the number of entry points and locations.

9. Determine electrical/mechanical/chemical hazards.

10. Define the hot, warm, and cold zones, and secure their perimeters.

Step Two: Personnel and Equipment

1. Ensure needed response of additional technical team members as required.

2. Ensure a full technical-rescue assignment.

3. Ensure adequate air supply, cascade truck, and bottles.

4. Ensure sufficient rehab area is established.

5. Ensure visible incident management and/or operations section is established.

6. Begin Tactical Work Sheet.

Step Three: Warm and Cold Zone

1. Establish a perimeter with tape and assign police to ensure an access point. Ensure that the battalion officer assigns an access control officer, preferably not a technical team member.

2. Ventilate the general area, if needed.

3. Ventilate the space. Continually assess the effectiveness of the ventilation process by (1) atmospheric monitor readings and (2) assessing the type and configuration of the space.

4. If possible, open all additional openings into the space to assist with the ventilation process; i.e., manholes, hatches, natural openings, etc.

5. Ensure fire control measures, if needed.

6. Do not allow sources of ignition on site.

Phase II: Entry Preparation

1. Ensure lockout, tag-out, blank-out procedures are complete.

    a. All fixed mechanical devices and equipment capable of causing injury shall be placed in a zero mechanical state (ZMS).

    b. All electrical equipment (excluding lighting) shall be locked out in the open (off) position with a key-type padlock.

    c. The key shall remain with the person who places his lock on the padlock.

    d. In cases where lockout isn't possible, equipment shall be properly tagged and physical security provided.

    e. All locked-out utilities shall also be tagged with an approved confined-space tag system.

2. Post nonessential personnel at those areas tagged and blanked or blinded.

3. Ensure that all personnel who will enter the site are equipped with SABA.

4. Ensure one backup team for every entry team. Assign RIT as necessary.

5. No one shall enter a confined space alone. Always work in teams.

6. Each entry team shall be equipped with the following items.

   a. One member shall have a sound-powered communications system in place, worn with the SABA, or radio.

   b. Hardwired package for communications.

   c. Explosionproof lighting.

   d. Atmospheric monitor, personal units preferred.

   e. Proper protective gear as deemed necessary by the incident manager. At the very least, each member shall wear fire-resistant coveralls, fire-resistant hood, boots, gloves, helmet, PAL, and respiratory protection.

   f. An entry/egress line shall accompany the first entry team and be anchored at their furthest point of penetration. If this line is equipped with a hardwired communications line (internal), it may function per section (b) as well.

   g. Some form of rapid extrication/retrieval harness for a victim.

   h. If the entry team must enter a vertical shaft of greater than five feet, each member shall wear a personal harness and be attached to a fall arresting system on entry.

   i. A victim SABA and supply line, if applicable.

Phase III: Atmospheric Monitoring

1. Atmospheric monitoring shall occur prior to and during all entries into a confined space.

2. Atmospheric monitoring should be accomplished at high and low areas of the space.

3. All atmospheres shall be tested for (a) pH, (b) oxygen deficiency, (c) oxygen enrichment, (d) flammability, and (e) toxicity.

4. The following levels shall be considered as IDLH environments:

   a. Oxygen deficient: <19.5%

   b. Oxygen enriched: >23.5%

   c. Flammability at 10% of LEL

   d. Toxicity shall be any limit whose numerical value exceeds the PEL, in accordance with the table.

5. Atmospheric monitoring shall occur during occupancy at intervals dependent on the possibility of changing conditions, but in no case less than hourly.

6. All atmospheric readings shall be recorded on a technical-rescue work sheet.

7. In the event that, in the opinion of the incident manager or his designee, the atmospheric readings become what he considers unsafe to continue operations, all entry teams shall be removed from the space immediately until such time as the atmospheric conditions are corrected.

Phase IV: Entry

1. Once the best method and location for entry has been determined, teams shall begin entry and reconnaissance/rescue/recovery operations in the space.

2. Entry decisions shall be made based on known locations of victims, the safety of the opening, atmospheric readings, and the ease of recovery points.

3. If possible, attempt a two-prong attack to reach the victim if his location is known or suspected.

4. Prior to entry, each team member shall be logged on a technical-rescue work sheet with his time of entry. This function shall be assigned to one technician, who shall keep the operations officer appraised of the status of each team.

5. Teams shall be limited to thirty minutes in the space.

6. Each team shall be assigned to rehab on removal from the space until rehydrated and vital signs are within normal limits.

7. Once inside the space:

    a. Ensure adequate interior team communications

    b. Ensure adequate communications with the operations exterior

    c. Mark with chalk, if necessary, the entry and movement patterns to ensure egress.

    d. Move toward the suspected victim location as a team.

    e. Beware of elevation differences and unstable footing.

8. Once the victim has been located, decide:

    a. Is this a rescue or recovery?

b. If rescue, can a SABA unit be placed on the victim?

c. Can the victim be easily moved toward the opening with the equipment carried by the team?

d. Is an additional team needed to make the move?

e. Communicate your decision to outside command.

9. Once the victim has been attached to a removal device and is in the process of being rescued/recovered, and if he/she is to be moved through an opening (either vertical or horizontal) that is the only route of egress, then the following guidelines are to be followed.

a. Whenever possible, ensure that all team members are stationed on the egress side of the opening in the event that the victim becomes lodged.

b. Always try to avoid being blocked in by the victim.

c. If this is not possible, ensure that (1) when the move is made, that it is made quickly and smoothly, leaving the time that the space is blocked for egress is minimal, (2) the exterior personnel, as well as the interior teams, are aware of the move and that a plan is agreed upon prior to blocking the space, and (3) all air lines and connections are clear of the victim and the path along which he will be moved.

Phase V: Victim Removal

1. Once the victim is set for removal, ensure the following.

a. Ensure as much C-spine control as possible, based on the space and the victim's condition.

b. Use removal systems on the exterior that are applicable to the size and weight of the victim.

c. Mechanical advantage systems are to be preferred over manual hauling.

    d. Do not use electric winches, etc., to remove victims.

    e. Decide whether the victim is to be removed headfirst or feetfirst.

    f. Avoid the use of wristlets on patients who have burns on their extremities.

2. Once the victim is clear of the space, remove all entry team personnel and equipment.

Phase VI: Safety Considerations

1. In the event of a air-line failure on a SABA, the entire team shall immediately leave the space, once they ensure that the rescuer with the problem is assisted.

    a. Notify the exterior team immediately of the problem and identify the line and the specific problem.

    b. Never leave a partner in trouble unless you must clear the way for his exit.

    c. In the event that the ten-minute bypass bottle runs out before you have exited and the air-line problem has not been corrected, (1) buddy breathing by passing the mainline (which is still functional) back and forth to each other's system is acceptable as a last-ditch survival effort, (2) do not leave the nonoperational line behind, and (3) exit the space and correct the problem.

Phase VII: Termination

1. Conduct PAR.

2. Have contractor or the responsible party seal entry points to ensure no additional entry.

3. Complete technical work sheet and documentation.

Reviewed by:

_____ Date _____
District chief, A shift

_____ Date _____
District chief, B shift

_____ Date _____
District chief, C shift

Approved as to content:

_____ Date _____
Deputy chief

Approved as to legal sufficiency:

_____ Date _____
City attorney

Approved:

_____ Date _____
Fire chief

Effective date of policy: _____

GLOSSARY OF
ATMOSPHERIC MONITORING TERMS

Accuracy: A meter that is exposed to a known concentration of gas and displays that amount is said to be accurate. If a monitor is exposed to a gas mixture containing 50 ppm of hydrogen sulfide several times and displays 50 ppm each time, the monitor is accurate. However, a meter can be accurate without being precise (see below).

Aerosol: A mixture, usually of liquids but sometimes solids, suspended in a gas. Spray paints and hair sprays are two common examples.

Alarm settings: A preset level within a monitor at which the monitor will display a visual alert and sound an audible alarm. Alarm settings are established by the manufacturer and based on OSHA and NIOSH levels for a given product. For confined space, we refer to these as action guidelines.

Bump test: Exposing a monitor to a known gas and allowing the monitor to go into alarm and then removing the gas is known as a bump test. This is also known as a field test.

Calibration: A monitor is calibrated to determine whether it is responding in an appropriate fashion. A full calibration is done by exposing the sensors to a known quantity of gas. New sensors will typically read higher than intended, and calibration electronically changes the sensor to read at the intended value. As a sensor gets older, it becomes less sensitive and calibration raises the value that the sensor displays, known as sensor span.

In accordance with 1910.146, your monitors should be calibrated according to the manufacturer's recommendations. This might be every month, every three months, or at an even greater interval. However, most manufacturers' literature requires calibration before each use. A full calibration during a rescue operation isn't feasible, and you should therefore use the bump test as a field check of the monitor. Regardless, teams should establish a regular schedule of calibration of every monitor and maintain a calibration log on each.

Combustible liquid: Liquids that have a flash point of 100°F or more.

Detection: The act of discovering the presence of a contaminant in a given atmosphere.

Detection range: This is a term used to express the unit of measure that the monitor uses to detect the vapor for which it was intended. Combustible gas indicators (CGI) usually have a display showing percent of the lower explosive limit (LEL). Toxic sensors of carbon monoxide and hydrogen sulfide display in parts per million (ppm).

Dusts: Solid materials that have been sanded, ground, or crushed.

Explosive limits: A reading or display on the monitor, given as a percentage, indicating a percentage of gas-in-air mixture, within which limits a given mixture will be combustible.

Field test: See *Bump test.*

Flammable liquid: A product with a flash point below 100°F (38.7°C).

Flammable range: The range of percentage of vapor in air that must be present to sustain combustion should ignition occur.

Flash point: The minimum temperature of a liquid that generates enough vapor to form an ignitable mixture in the vapor space above the liquid.

Ignition temperature: The minimum temperature to which a liquid must be raised to initiate and sustain combustion.

Immediately Dangerous to Life and Health (IDLH): The maximum concentration of a vapor from which a person could escape, in the event of respirator failure, without permanent or escape-impairing effects within thirty minutes.

Lag time: The amount of time it takes the air sample to reach the sensor's receiving area. In general, the value of one second per foot of sampling hose can be used for lag time. Most monitors have a three- to fifteen-second lag time. You should be familiar with the lag time for your specific monitor, since it depends on the length of hose, the type of pump system, and the capacity of the pump on the monitor.

Lower explosive limit (LEL): The minimum concentration of a vapor in air at which propagation of flame occurs on contact with a source of ignition. It is usually expressed as a percentage of gas vapor in air.

Mists: Liquids suspended in gas.

Monitoring: The act of measuring the amount of a contaminant at a given location and at a given time.

Oxygen sensor: An electrochemical sealed unit that measures the percentage of oxygen in the air. The sensor has two electrodes, an electrolyte solution, and a membrane that separates the two. As O_2 passes through the membrane, a reaction with the solution and the electrodes produces an electrical current that causes the sensor to display the percent of O_2 found.

Permissible exposure limit (PEL): The average concentration that must not be exceeded during an eight-hour shift or an forty-hour workweek. This average includes a one-hour clean-air break.

Precision: This term is used to describe the ability of a monitor to reproduce the same results each time it samples the same atmosphere. If a monitor is exposed to 100 ppm of carbon monoxide, it should display 100 ppm. If the monitor displays 50 ppm each time, it is still precise, it just isn't accurate. A meter that displays 70, 85, and 90 for three readings of the same sample is more accurate, since its results are closer to the actual 100 ppm present, but it isn't as precise as the other meter, since it displays a variety of readings.

Reaction time: The amount of time that a meter (actually, the sensors) requires to interpret the incoming sample and display a value on the screen. Also known as response time.

Recovery time: The time that it takes a monitor (actually, the sensors) to clear itself of a given sample. You must keep in mind the reaction time and lag time of your particular monitor, since this will affect the recovery time.

Relative response: Used to describe the way a monitor reacts to a gas other than the one it was intended to find. All monitors are designed to be used for one specific gas, and are calibrated to that gas. This makes the monitor accurate only for that gas; still, the monitor may be used for detecting others.

Relative response curve: A mathematical expression, in graphic form, of a monitor's ability to respond to other gases.

Short-term exposure limit (STEL): The fifteen-minute exposure limit that must not be exceeded during a workday.

Threshold limit value/time-weighted average (TLV/TWA): The average concentration limit for a normal eight-hour workday and a forty-hour workweek that should not cause adverse effects.

Threshold value limit-ceiling: The concentration that should never be exceeded.

Upper explosive limit (UEL): The maximum concentration of a vapor in air, at which propagation of flame occurs in the presence of a source of ignition.

Warmup time: Every electronic instrument, including a monitor, requires some time to warm up. An average warmup time for electronic monitors is about two minutes. Knowing the warmup time is important to ensure that you're getting accurate readings.

SUMMARY OF DIVISION AND ZONE CLASSIFICATION SYSTEM

Class I, Division 1

A location in which ignitable concentrations of flammable gases, vapors, or liquids:

- can exist under normal operating conditions,

- may exist frequently because of repair or maintenance operations or because of leakage, or

- may exist because of equipment breakdown that simultaneously causes the equipment to become a source of ignition.

The equipment intended for use in a Class I, Division 1 areas is usually of the explosionproof, intrinsically safe, or purged/pressurized variety.

Class I, Division 2

This type of environment is one in which:
- volatile flammable liquids or flammable gases or vapors exist, but are normally confined within closed containers,

- ignitable concentrations of gases, vapors, or liquids are normally prevented by positive mechanical ventilation, or

- the space is adjacent to a Class I, Division 1 location where ignitable concentrations might occasionally be communicated.

Equipment intended for use in a Class I, Division 2 area is usually of the nonincendive, nonsparking, purged/pressurized, hermetically sealed, or sealed-device variety.

Class I, Zone 1

A Class I, Zone 1 location is one in which ignitable concentrations of flammable gases, vapors, or liquids:

- are likely to exist under normal operating conditions,

- may exist frequently because of repair or maintenance operations or leakage,

- may exist because of equipment breakdown that simultaneously causes the equipment to become a source of ignition, or

- are adjacent to a Class I, Zone 0 location from which ignitable concentrations could be communicated.

Equipment intended for use in a Class I, Zone 1 area is usually of the flameproof, purged/pressurized, oil-immersed, increased-safety, encapsulated, powder-filled variety.

Class I, Zone 2

This classification of environment is one in which:
- ignitable concentrations of flammable gases, vapors, or liquids aren't likely to occur in normal operation or, if they do, will exist only for a short period,

- volatile flammable liquids, gases, or vapors exist but are normally within closed containers,

- ignitable concentrations of gases, vapors, or liquids are normally prevented by positive-pressure ventilation, or

- the location is adjacent to a Class I, Zone 1 location from which ignitable concentrations could be communicated.

The equipment intended for use in a Class I, Zone 2 area is usually of the nonincendive, nonsparking, restricted-breathing, hermetically sealed, or sealed-device variety.

Class II, Division 1

This type of environment is one in which:

- ignitable concentrations of combustible dust can exist in the air under normal operating conditions,

- ignitable concentrations of combustible dusts may exist because of equipment breakdown that simultaneously causes the equipment to become a source of ignition, or

- electronically conductive combustible dusts may be present in hazardous quantities.

Equipment intended for use in this category of environment is usually of the dust-ignitionproof, intrinsically safe, or pressurized type.

Class II, Division 2

This sort of location is one in which:

- a combustible is not normally in the air in ignitable concentrations,

- dust accumulations are normally insufficient to interfere with normal operation of electrical equipment,

- dust may be in suspension in the air as a result of infrequent malfunctioning of equipment, or

- dust accumulation may be sufficient to interfere with safe dissipation or may be ignitable by abnormal operation.

Equipment intended for use in this type of environment is usually of the dust-tight, nonincendive, nonsparking, or pressurized variety.

Class III, Division 1

A Class III, Division 1 environment is a location in which easily ignitable fibers or materials producing combustible particles are handled, manufactured, or used. Equipment intended for use in such a location is usually of the dust-tight or intrinsically safe variety.

Class III, Division 2

A Class III, Division 2 location is one in which easily ignitable fibers are stored or handled. Equipment intended for use in such an area is usually of the dust-tight or intrinsically safe variety.

INCIDENT ACTION PLAN

Incident name: _____ Date: _____

First IC: _____

Operational period (date/time): _____

    Strategic goals:

    Tactical objectives:

    Results:

Second IC: _____

Operational period (date/time): _____

    Strategic goals:

    Tactical objectives:

    Results:

Third IC: _____

Operational period (date/time): _____

    Strategic goals:

    Tactical objectives:

    Results:

POSTINCIDENT ANALYSIS

Address: _____

Unit: _____

Date: _____ Time of Arrival: _____

Nature of Incident: _____

Describe the situation on arrival.

Describe the assignments given and actions taken in chronological order.

A. Given by: _____ Given to: _____
    Actions taken/observed:

B. Given by: _____ Given to: _____
    Actions taken/observed:

C. Given by: _____ Given to: _____
    Actions taken/observed:

Obstacles encountered (note comments on back of page)

_____ Communications _____ Coordination _____ Equipment

_____ Management _____ Personnel _____ Safety

_____ Other, please specify: _____

_____

POSTINCIDENT ANALYSIS SUMMARY

Date: _____   Time of Alarm: _____

Address: _____

Type of incident: _____

Situation on arrival of first companies. Include a brief description
of the situation encountered by the first companies on the scene.

Outcome of incident. List the extent of damage to the original
occupancy and exposures. Include damage to fire equipment, as
well as civilian and personnel casualties.

Equipment committed to the incident. List the resources committed to the incident, those stations backfilled (by which apparatus), and those stations (if any) left empty.

Answers to Chapter Questions

CHAPTER ONE

Toward Forming a Confined-Space Team

1. True.

2. NFPA 1983.

3. NFPA 1670.

4. Technical rescue requires personnel, equipment, and training not traditionally mandated by conventional fire and rescue services.

5. By establishing practical standards.

6. When a call comes in, personnel will invariably deploy to the emergency environment, even if the members sent in are wholly untrained for that sort of incident.

7. Team effort.

8. Abstract qualities.

9. Capability, performance, risk factors, and chances for success.

10. To build a constituency to support the team.

CHAPTER TWO
Managing the Risks

1. (1) People, (2) policy, (3) training, (4) supervision, and (5) discipline.

2. 35 percent.

3. 4.8 million.

4. A confined space (1) is large enough and configured so that an employee can enter it, (2) has limited means of egress, and (3) is not designed for continuous occupancy.

5. A space that an employer can document as being hazard-free for as long as any worker is inside.

6. Asphyxiation due to a hazardous atmosphere.

7. Because employees are poorly trained and equipped to handle the environment that they encounter.

8. The agency does not intend to specify exact methods of compliance, provided that the employer's program results in worker performance that meets the requirements and prevents unsafe acts.

9. NFPA 1670 codifies training requirements, organizational planning, definitions, and the caliber of operations for technical rescue.

226

10. (1) Structural collapse, (2) rope rescue, (3) confined-space rescue, (4) vehicle and machinery rescue, (5) water rescue, (6) wilderness search and rescue, and (7) trench rescue.

11. Yes.

12. No.

13. (1) The internal configuration of the space is clear and unobstructed, so that retrieval systems can be used for rescuers without chance of entanglement, (2) the victim can easily be seen through the primary access opening to the space, (3) the rescuer can easily pass through the access opening while wearing PPE, (4) the space can accommodate two or more rescuers, in addition to the victim, and (5) all hazards in and around the space have been identified, isolated, and controlled.

CHAPTER THREE
Planning

1. Because it is unlikely that the only confined spaces in the juris-diction are connected with industries that maintain prepared brigades. Thus, the organization will sooner or later respond to incidents at which it must act as the primary agency.

2. (1) Assessment of the hazards and risks, (2) development of internal and external resources, (3) assessment of team readiness, (4) training of the team, and (5) recurrent training and periodic reassessment of the program.

3. True.

4. External resources.

5. If you can show that you attempted to acquire those resources and were denied for budgetary reasons, then, in the event of a mishap, the burden of explanation or liability may fall elsewhere.

6. In keeping the books on any donations or income, an independent body will face any audits rather than an individual or the team itself. This arrangement also provides a tax write-off opportunity for companies and individuals who donate to the team.

7. The team's most effective member.

CHAPTER FOUR
Training the Team

1. Risk.

2. Development of the team.

3. In the classroom.

4. (1) To foster teamwork, (2) to have a student work with a particular piece of equipment in a confined space, (3) to prepare the student psychologically for real-world confined spaces, and (4) to build on basic skills.

5. By pushing the students to a point where they are working so quickly that they become inefficient, the instructor can then slow them down again, demonstrating how they can actually increase their proficiency and cooperation.

6. Team members are required to place teams of two inside a vertical space, then remove them using a variety of methods.

7. Teams must operate exactly as they would in the field, using all of the equipment and other resources at their disposal.

8. In a managed exercise, the instructor actively coaches the operation. In a facilitated exercise, no feedback is provided until the end, and then only in the form of a critique.

9. Awareness, Operations, Technician, and instructor.

10. At least once per quarter and for an entire day.

CHAPTER FIVE
Equipment

1. Division system.

2. Zone system.

3. (1) Explosionproof, (2) intrinsically safe, and (3) purged device.

4. Class, division, and group.

5. No.

6. That the equipment has been approved by the Canadian Standards Association, and that OSHA has approved the equipment for use in the United States, based on adherence to test methods specified by UL and ANSI.

7. (1) Air-purifying respirator, (2) SCBA, (3) rebreather, and (4) SABA.

8. (1) Electrocatalytic, (2) semiconductor, (3) electrochemcial, (4) infrared, and (5) photo-ionization.

9. 3,000 cfm.

10. One tug meant *Okay,* two meant *Advance,* three meant *Take up,* and four meant *Help.*

11. A wireless system is subject to dead spots and intermittent communications, and it is not a hands-free system, since it requires the user to push a mike button to talk.

12. Nylon kernmantle and polyester.

13. Five feet.

14. The same individual who placed it there.

15. Only in a last-ditch attempt to save someone's life.

CHAPTER SIX
Commanding the Incident

1. To use incident management every day, on every call to which you respond.

2. Rescue operations officer.

3. Extrication officer.

4. Entry team officer.

5. A technical rescue physician or a senior technical-rescue paramedic.

6. Rig master.

7. Supply officer.

CHAPTER SEVEN
The Initial Response

1. Immediately on arrival.

2. Since an incident that turns out to be a confined-space rescue may, in fact, have been dispatched as an illness call or report of a man down, a resource assessment is vital toward ensuring that the proper units respond.

3. Good.

4. (1) An immediate or delayed threat to life, (2) a threat that would cause irreversible adverse health effects, and (3) a threat that would interfere with an individual's ability to escape from the space without assistance.

5. (1) Atmospheric hazards, (2) burn hazards, (3) mechanical hazards, (4) engulfment hazards, and (5) hazardous materials.

6. 19.5 percent O_2.

7. 23.5 percent.

8. Hydrogen sulfide and carbon monoxide.

9. Permissible exposure limit.

10. Eight hours.

11. Thermal, Radiation, Asphyxiation, Chemical, Etiologic, Mechanical.

12. Chief engineer.

CHAPTER EIGHT

Operations Prior to Entry

1. Never.

2. Rendering a length of pipe safe by locking out and tagging a vent valve located between two other in-line valves that have been closed off.

3. A depiction of the response of a monitor to gases other than the one that it was primarily designed to detect.

4. Vapor density.

5. The tendency of a liquid to evaporate.

6. (1) pH, (2) oxygen, (3) flammability, (4) toxicity.

7. True.

8. Organics.

9. 10 percent.

10. Ten to fifteen.

11. False.

12. The operations officer in conjunction with the extrication and safety officers.

CHAPTER NINE
Entry and Rescue Operations

1. False.

2. False.

3. Odd.

4. Putting all of the links to the outside world in a single case, usually a piece of two-inch tubular webbing.

5. Two.

6. Doing so would leave nothing for the next team trying to retrace his route inward.

7. True.

8. False.

CHAPTER TEN
Removing the Victim

1. On the egress side of him.

2. True.

3. There is a tendency for exterior teams to become complacent, since they aren't yet directly involved in handling the victim.

4. Because your inability to judge resistance may result in severe injuries to the victim.

5. False.

6. True.

7. True.

8. False.

CHAPTER ELEVEN

Termination Activities

1. The operations officer.

2. According to a plan developed through the incident management system and codified in the SOPs.

3. Personnel accountability system.

4. OSHA.

5. Certain personnel are critical to the postoperation and must remain.

6. Document the name, address, telephone number, and title of the responsible party. Note the exact time that you turned over the site. Document that you provided the responsible party with any safety information regarding how the site was left, including lockout, tag-out information. Restrict access to the space, either by locking it off or stretching fire line tape around the general area.

7. To help you evaluate whether your systems and devices are holding up under the rigors of real-world operations; i.e., to pinpoint any misuse, flaws in design, and main-tenance concerns.

8. The incident commander.

INDEX

F

G

H

I

N

O

S

T

U